AF782980

Knowledge Engineering

Knowledge Engineering

The Process Paradigm

Hamed Fazlollahtabar

Department of Industrial Engineering, School of Engineering,
Damghan University, Damghan, Iran

CRC Press
Taylor & Francis Group
Boca Raton London New York

CRC Press is an imprint of the
Taylor & Francis Group, an **informa** business

First edition published 2020
by CRC Press
6000 Broken Sound Parkway NW, Suite 300, Boca Raton, FL 33487-2742

and by CRC Press
2 Park Square, Milton Park, Abingdon, Oxon, OX14 4RN

CRC Press is an imprint of Taylor & Francis Group, LLC

ISBN: 978-0-367-51736-6 (hbk)
ISBN: 978-1-003-05500-6 (ebk)

Typeset in Times
by Deanta Global Publishing Services, Chennai, India

Contents

Preface....ix
Introduction....xi

Chapter 1 Knowledge and Global Learning....1
Main Body....1
Subsidiaries' Knowledge Stock....1
Knowledge Transfer Mechanisms....1
Data and Methods....2
Dependent Variables....2
Independent Variables....2
Control Variables....3
Methods....3
Assessment of Measures....4
Concluding Remarks....4
References....5

Chapter 2 Knowledge Processes, Intensity and Innovation....7
Introduction....7
Literature Review....7
Main Body....8
Hypotheses....8
Analysis....9
Measurement....12
Concluding Remarks....15
References....17

Chapter 3 Knowledge with Innovation Performance....19
Introduction....19
Main Body....19
Properties of Knowledge....20
Absorptive Capacity....21
The Research Framework of the Study....22
Methods and Measurement....22
Data Collection and the Sample....22
Definitions and Measurements of the Constructs....23
Dependent Variables....23
Independent Variables....23
Moderate Variables....23
Results....24

Model and Analysis 24
Regression Analysis of Innovation Performance on Properties of Knowledge 25
The Moderate Effects of a Firm's Absorptive Capacity 25
Concluding Remarks 27
References 31

Chapter 4 Knowledge and Innovation in Networked Environments 33

Literature Review 33
From Dynamic Capabilities to Knowledge-Based Dynamic Capabilities 33
The Construct of Knowledge-Based Dynamic Capabilities 33
Network Embeddedness and Innovation 35
Hypotheses Development 35
The Link between Knowledge-Based Dynamic Capabilities and Innovation 35
The Antecedents of Dynamic Capabilities: Network Embeddedness 35
Methods and Materials 37
Dependent Variable: Innovation Performance 37
Knowledge-Based Dynamic Capabilities (KDC) 37
Independent Variables: Network Embeddedness 38
Evaluations 40
Construct Validity and Reliability 40
Hypothesis Testing 40
Robustness of the Results 41
Concluding Remarks 41
References 42

Chapter 5 Knowledge and Organizational Business Loss 45

Research Scope 45
Literature Review 45
Key Success Factors Described in Literature 47
Literature Critique 47
Lessons Learned 47
Proposed Framework 47
Research Method and Results 48
Concluding Remarks 52
References and Further Reading 52

Chapter 6 Knowledge Sharing Using Semantic Web 55

Introduction 55
Main Body 56
Articulation 58

Knowledge-Sharing Platform 58
Concluding Remarks 62
References 63

Chapter 7 Knowledge Sharing and Tax Payment 65

Introduction and Background 65
Proposed Problem and Model 68
Risk Modeling Structure 70
Interval Programming 71
Case Study 72
Computing Mean Values (t_i) 73
Computing Loss Function (L_i) 75
Computing $f(x_i)$ for the Factors 76
Computing Risk Function ($R(\underline{x}_i)$) 78
Concluding Remarks 79
References 80

Chapter 8 Knowledge Sharing for Enterprise Resources 83

Introduction and Background 83
Proposed Framework 85
Network Range 87
Tie Strength 88
Network Efficiency 89
Measuring CNIAM 89
Case Study 89
Concluding Remarks 92
References 92

Chapter 9 Knowledge Sharing and Organizational Culture 93

Introduction 93
Literature Review 94
Research Importance and Necessity 100
Research Hypotheses 100
Research Method 101
Data Analysis 102
Variable Rankings Based on Friedman Test 108
Concluding Remarks 109
References 111

Chapter 10 Knowledge Sharing and Learning Capability 113

Introduction 113
Organizational Learning Capability 114
Organizational Learning Capability Dimensions 114

Risk-Taking 114
Interaction with the External Environment 115
Dialogue 115
Participative Decision Making 115
Managerial Commitment 115
Systems Perspective 116
Openness and Experimentation 116
Knowledge Transfer and Integration 116
Teamwork 117
Demonstration of Mission and Goals 117
Problem Definition and Modeling 117
Computational Results 121
Concluding Remarks 121
References 122

Index 125

Preface

Knowledge management is far-reaching. Maybe you are considering developing your own knowledge management competencies to become a more effective player in the global knowledge economy, or to become a more competitive knowledge leader and knowledge-driven organization. Maybe you want to develop and apply knowledge management strategies to a government, military operations, global poverty eradication and international disaster management and even, now, knowledge management for global climate change. The list is endless. Knowledge management is applied today across the world, in all industry sectors, public and private organizations, humanitarian institutions and international charities. Most importantly, effective knowledge management is now recognized to be "the key driver of new knowledge and new ideas" to the innovation process, to new innovative products, services and solutions.

Effective knowledge management should dramatically reduce costs such as those of office work repetition, human resource retirement, information reuse, etc. Most individuals, teams and organizations are today continually "reinventing the wheel"! It might be that they simply do not know if something has already been done. They might not know what is already known, or they do not know where and how to access the knowledge. Continually reinventing the wheel is a costly and inefficient activity, whereas more systematic reuse of knowledge will show substantial cost benefits immediately. Apart from reducing the costs, effective knowledge management should also increase our speed of response as a direct result of better knowledge access and application. Effective knowledge management, using more collective and systematic processes, will also reduce our tendency to "repeat the same mistakes"! This is, again, extremely costly and inefficient. Effective knowledge management, therefore, can dramatically improve the quality of products and services. Better knowing our stakeholder needs, customer needs, employee needs and industry needs, for example, has an obvious immediate effect on our relationship management.

This book is composed of ten chapters, focusing on different processes in knowledge management, in the engineering model, to effectively develop all aspects of knowledge management in the paradigm of knowledge innovation and economy. The first chapter includes global learning processes and knowledge concepts. In Chapter 2, innovation and knowledge management relations are studied. Chapter 3 contains performance management in knowledge processes. In Chapter 4, a network structure of innovation and the knowledge process are considered, being a significant model for the knowledge economy. Chapter 5 presents the organizational place of knowledge management. In Chapters 6 to 10, different aspects and process models of knowledge sharing are represented. Knowledge sharing is a significant concept and very challenging in knowledge engineering. This way, Chapter 6 models knowledge sharing using a semantic web approach. Chapter 7 concentrates on the risk between knowledge sharing and tax payment in the economy. Chapter 8 investigates a strategic model and evaluation process for the enterprise knowledge sharing process. Chapter 9 analyzes organizational culture and knowledge sharing in industry.

Chapter 10 models learning capability and knowledge sharing mathematically as an engineering paradigm.

The audience will make use of very practical models that can be implemented. In addition to that, academicians can find analytical models of knowledge engineering, applicable to different environments.

Academicians and practitioners in the area of knowledge management and engineering, especially managers in industry, can be the target audience. Also, the material could be used as a textbook for knowledge management courses in graduate studies.

Hamed Fazlollahtabar
Department of Industrial Engineering
School of Engineering
Damghan University
Damghan, Iran

Introduction

Knowledge management is essential in the new era of knowledge-driven systems and processes. Among them, knowledge engineering for enterprises concerning process concept is developing drastically. This book is for knowledge management experts and practitioners. It is also effective for those who seek organizational influences and analytical perspectives of knowledge in enterprises. The purpose of this book is to provide an introduction to the various process-oriented knowledge engineering in enterprises. Knowledge sharing is a significant concept and very challenging in knowledge engineering. It is meant for a wide audience of readers interested in gaining some understanding of the basics of knowledge management and engineering and at the same time their applications in real cases. These include professionals in the private sector, managers of corporations and business executives, as well as government officials at various levels who may have a management, technical or engineering background but no exposure yet to knowledge management.

This book has ten chapters. The first chapter discusses knowledge management within the global learning process concept. In Chapter 2, innovation and knowledge management relations are investigated. Chapter 3 contains performance management in knowledge processes. In Chapter 4, a network structure of innovation and a knowledge process are considered. Chapter 5 presents the organizational perspective of knowledge management. In Chapter 6, knowledge sharing using a semantic web approach is modeled. Chapter 7 focuses on the risk between knowledge sharing and tax payments. Chapter 8 develops a knowledge sharing process evaluation model. Chapter 9 analyses organizational culture and knowledge sharing in industry. Chapter 10 provides a mathematical model for learning capability and knowledge sharing.

I recommend this book for all academicians, researchers and practitioners in the field of knowledge management.

1 Knowledge and Global Learning

MAIN BODY

Subsidiaries' Knowledge Stock

For foreign subsidiaries, a major source of learning is their operational experience in the host country. The subsidiary's host country experience subdivides into a time-based experience and a diversity-based experience. The longer the time-based experience and the more diverse the experience of the subsidiary in the host country, the more knowledge is transferred from it to the parent company and to its peer subsidiaries [1,2].

Hypotheses can be expressed as follows:

H1a/H1b: The longer the time-based experience of the subsidiary in the host country, the more knowledge is transferred to the parent company (H1a) and to its peer subsidiaries (H1b).

Centralized control imposed by a parent company undermines a subsidiary's motivation to learn, thus the more autonomy is given to the subsidiary, the more knowledge is transferred to the parent company and to its peer subsidiaries.

H1c/H1d: The more diverse the experience of the subsidiary in the host country, the more knowledge is transferred to the parent company (H1c) and to its peer subsidiaries (H1d).

Knowledge Transfer Mechanisms

Poor communication and lack of incentives could hinder the transfer and utilization of the knowledge. Staffing expatriate managers in foreign subsidiaries and facilitating direct contact between managers in different subunits can be effective. Expatriate managers can share work experience and culture with parent company managers [3–5].

H3a/H3b: The more expatriates there are in the subsidiary, the more knowledge is transferred to the parent company (H3a) and to its peer subsidiaries (H3b).

Frequent communication among staff has a crucial role in the transfer of innovations.

H4a/H4b: The more frequent the communication between subsidiary managers and parent company managers, the more knowledge is transferred from the subsidiary to the parent company (H4a) and to its peer subsidiaries (H4b).

DATA AND METHODS

Data were collected through an email survey of foreign subsidiaries operating in Iran. First, all firms engaged in manufacturing activities and which had at least two years of operational experience were identified. Then, firms whose employee base was less than 50 were excluded, and firms in which foreign parent firms had less than 50% equity stake were also excluded. In total, 404 firms were included in the email survey. Next, the first version of the questionnaire was prepared and pre-tested by three senior managers from different foreign-owned firms. The questionnaire was revised and finalized according to their comments. Then, they explained the purpose of the study and asked if the team member was willing to participate in the email survey. Managers in 195 out of 404 firms agreed to do so. Finally, 98 completed questionnaires were received. Among these, 17 questionnaires, which had insufficient responses (four cases) or reported conducting only manufacturing activities but no sales activities (13 cases) were excluded. Of the remaining 81, ten respondents noted that their parent companies had no foreign affiliates other than the Iranian subsidiary. Therefore, 81 cases were available for analyzing knowledge transfer to the parent company, and 71 cases to peer subsidiaries.

Dependent Variables

First, dependent variables measured the amount of knowledge that the focal subsidiary transferred to the parent company in five major dimensions:

1. The development of basic and applied technologies;
2. New product design and development;
3. Manufacturing activities;
4. Sales, marketing and distribution; and
5. General management (Cronbach's $\alpha = 0.911$).

The second dependent variable was measured as the amount of knowledge that the focal subsidiary transferred to peer subsidiaries (Cronbach's $\alpha = 0.941$).

Independent Variables

The time-based experience was measured by the number of operational years of the subsidiary. The diversity-based experience was operationalized by the extent to which the subsidiary:

- Produced and sold diverse products;
- Served diverse wholesale markets;

- Served diverse retail markets; and
- Dealt with diverse buyers and customers (seven-point Likert scale).

Subsidiary autonomy was measured, by asking the respondents to indicate the extent to which a subsidiary could make its own decisions without interference from the parent company, in six management domains:

1. Development and launch of new products;
2. Pricing decisions and marketing activities;
3. Expansion and reduction of manufacturing facilities;
4. Human resources management policies;
5. Borrowing and raising capital; and
6. Setting annual business goals (seven-point Likert scale).

The expatriate policy for the basic subsidiary was measured at three different levels using three different variables:

- Dummy variable;
- The number of expatriates; and
- The ratio of expatriates to the total number of employees in the subsidiary.

Communication frequency between the parent and subsidiary managers was measured according to:

- How often the subsidiary manager communicates with managers in the parent company through e-mail, telephone, etc.;
- How often the subsidiary manager makes business trips to the parent company to have face-to-face meetings.

Control Variables

- The size of the subsidiary: measured by the total number of employees;
- The ownership percentage of the foreign parent in the subsidiary: measured by a dummy variable, if the value is "1" the foreign parent had an equity stake over 95%, and "0" otherwise.

Methods

Ordinary Least Square (OLS) regression was used to test the hypotheses.

The regression is performed on:

- Two dependent variables (knowledge transfer to a parent company and knowledge transfer to peer subsidiaries);
- Five independent variables (subsidiary experience, subsidiary autonomy, expatriate policy, communication frequency and performance appraisal);

- Seven control variables (subsidiary size, parent ownership, three industrial dummies and two home country dummies).

Assessment of Measures

The authors first performed exploratory factor analysis for four multi-item scales (i.e., diversity of experience, subsidiary autonomy, frequency of communication with a parent, frequency of communication with peers). Each of the constructs is confirmed as one dimension, with the exception of subsidiary autonomy, whose related items were grouped into two different factors: one associated with activities such as the expansion and reduction of manufacturing facilities, and the other, a confirmatory factor analysis, was run to evaluate the convergent and discriminant validity of the five factors extracted by exploratory factor analysis. Results show a good fit for the five-factor model ($\chi^2 = 73.63$, $df = 67$, $P = 0.38$, comparative fit index = CFI = 0.98, goodness-of-fit index = GFI = 0.89, incremental fit index = IFI = 0.98, root mean square error of approximation = RMSEA = 0.023).

All of the items loaded significantly on their latent variable (*t*-statistics ranging from 3.91 to 9.62), indicating the convergent validity of the measures. To check discriminant validity, the author compared the model, in which the latent variables were allowed to correlate freely, with a series of nested models in which each pair of constructs was restricted to correlate perfectly. All the chi-square differences are highly significant ($p = 0.000$), supporting discriminant validity. Further, the author compared the five-factor model with the alternative four-factor model, which includes only a single construct for autonomy. Fit index comparison for the first and second models ($\chi^2 = 94.97$, $df = 71$, P = 0.03, RMSEA = 0.061, CFI = 0.95, GFI = 0.86) and the chi-square difference test reveal that the five-factor model fits the data better. Thus, this study measured subsidiary autonomy in two separate dimensions, labeling them autonomy 1 (production capacity expansion, capital raising and business goal setting) and autonomy 2 (new product introduction, marketing and human resources management (HRM)), respectively. Also, Cronbach's alphas for all of these independent variables range from 0.660 to 0.828, suggesting adequate internal consistency. Lastly, the author checked the common methods variance of the analysis using Harman's one-factor test, since the independent and dependent variables were collected from single respondents. This test resulted in four factors with eigenvalues above 1, with the largest factor accounting for 28.2% of the variance in the sample. Therefore, common methods variance does not appear to be a problem in this empirical analysis.

CONCLUDING REMARKS

Transferring knowledge between a parent company and its peer subsidiaries is different. Although subsidiary skills and experience have increasingly been viewed as an important source of new competitive advantage, the analysis suggests that it is challenging for multinational corporations (MNCs) to nurture and utilize the subsidiary capabilities effectively for two main reasons [6–8]. First, although there is evidence that the accumulation of new knowledge at the subsidiary level is an important condition for efficient global learning, the parent company often needs to exert strong control over foreign subsidiaries to create synergies and leverage inter-unit interdependencies. Thus, the ability of MNCs to encourage foreign subsidiaries to experiment and come up with new ideas and solutions on their own to create new knowledge can be constrained by the parent's strategic need to integrate subsidiary activities. Second, the findings of this study highlight the difficulty of transferring a subsidiary's knowledge to peer subsidiaries due to the lack of absorptive capacity and formal communication channels among subsidiaries [9,10]. These conditions may prevent a foreign subsidiary from efficiently sharing its best practices and skills with other subsidiaries in the same or neighboring region, even when the subsidiary has developed new and valuable skills, which are readily applicable to its peers owing to geographical proximity [11,12].

Transferring the knowledge of a subsidiary to the parent company is related to:

- The diversity of experience;
- First autonomy (such as the expansion and reduction of manufacturing facilities, borrowing and raising capital and setting annual business goals);
- Second autonomy (such as development and launch of a new product, pricing decisions, marketing activities and HRM policies);
- The number of expatriates and evaluation of the subsidiary company;
- Transferring the knowledge of the subsidiary to its peer subsidiaries is related to the time-based experience;
- Communication and evaluation of the subsidiary company.

REFERENCES

1. Harzing A. Acquisitions versus greenfield investments: international strategy and management of entry modes. *Strat Manag J.* 2002;23(3):211–27.
2. Rugman AM, Verbeke A. A perspective on regional and global strategies of multinational enterprises. *J Int Bus Stud.* 2004;35(1):3–18.
3. Song J, Shin J. The Paradox of technological capabilities: a study of knowledge sourcing from host countries of overseas R&D operations. *J Int Bus Stud.* 2008;39(2):291–303.
4. Crowne KA. Enhancing knowledge transfer during and after international assignments. *J Knowl Manag.* 2009;13(4):134–47.
5. Monteiro LF, Arvidsson N, Birkinshaw J. Knowledge flows within multinational corporations: explaining subsidiary isolation and its performance implications. *Org Sci.* 2008;19(1):90–107.
6. Amsden AH. *The rise of "the rest": challenges to the west from late-industrializing economies.* New York, NY: Oxford University Press; 2003.
7. Cantwell J, Mudambi R. MNE competence-creating subsidiary mandates. *Strat Manag J.* 2005;26(12):1109–28.
8. Noorderhaven N, Harzing A. Knowledge-sharing and social interaction within MNEs. *J Int Bus Stud.* 2009;40(5):719–41.
9. Lazarova M, Tarique I. Knowledge transfer upon repatriation. *J World Bus.* 2005;40(4):361–73.
10. Corredoira RA, Rosenkopf L. Should auld acquaintance be forgot? The reverse transfers of knowledge through mobility ties. *Strat Manag J.* 2010;31(2):159–81.

11. Schleimer S, Riege A. Knowledge transfer between globally dispersed units at BMW. *J Knowl Manag.* 2009;13(1):27–41.
12. Fazlollahtabar H, Aghasi E. *Knowledge management: from concept to engineering implementation.* Mashhad, Iran: Sabet Ghadam Publishing; 2012. ISBN: 978-600-91153-8-9 (in Persian).

2 Knowledge Processes, Intensity and Innovation

INTRODUCTION

Nowadays, innovation and innovativeness are important in every organization. On the other hand, one of the most effective things in innovation is knowledge management. In this chapter, we want to elaborate on knowledge management and innovation and explore the relationship between them. In the first step, we will specify knowledge management models having an acceptable impact on innovation. A mass data collection from three provinces of Iran was conducted, and the regression method was considered as a way to find the answer to the "how many" and "what is" relationships between the variables and innovation.

Economists believe that innovation is a competitive advantage that can be a basis for sustainable development in the current knowledge economy. Generally, innovation is a process that can create an idea and then implemented to create value [1]. The focal point in innovation literature is effective for knowledge management [2–5], but in implementation of the ideas are very general with no details. As a matter of fact, what part of knowledge model has meaning and has a better effect on innovation? In the literature, most of the works are focused on the relationship between innovation and one or two components of models [6]. A few works considered all parts of a particular model [2]. Most studies tried to work on the direct and flat model without intermediate assumption. This is a gap that is handled in this chapter using empirical analysis of some organizations that are active in knowledge management. In addition, a variable that has a profound effect on every knowledge structure is knowledge intensiveness. The structure of this chapter is as follows: in the first step, we will have a review of the literature. Next, the research strategy is presented, including data collection methods and sample characteristics. Then, the research findings are reported; and finally, the chapter concludes with theoretical achievements and managerial implications.

LITERATURE REVIEW

Some works believe that the innovation process is independent of knowledge management, whereas some researchers concluded that knowledge creation in knowledge management influences innovation. There is not any generic model that can describe every process and component of innovation. A key premise in the literature on new product innovation is that the rate of new product introduction is a function of a firm's ability to manage, maintain and create knowledge. Some case studies in the literature emphasized that innovation and knowledge management are interdependent. However, there is lack of quantitative research for analytical investigations. Generally, understanding the process of knowledge management and its relationship with innovation is still doubtful. As past works imply, innovation is the outcome of the knowledge management process. The literature typically identifies four or six processes [7], but we can choose an integrated model adopted in the literature with these components: documentation, knowledge sharing, knowledge creation and knowledge acquisition.

Documentation refers to coding and registering any knowledge that is in the company or organization. Knowledge sharing refers to sharing every acquisition and all saved knowledge in the organization. Knowledge creation is producing and burning any knowledge in the company. The last component is knowledge acquisition, referring to taking any knowledge out of (external) or into (internal) the company. Based on the proposed model, we have an uncoordinated image in the literature about the impact of knowledge on innovation. For example, [8] presented empirical evidence that both external knowledge acquisition and internal knowledge sharing will increase the innovativeness of every company and said that it is external knowledge acquisition leading to organizational innovation. Chou [6] showed that knowledge acquisition has an effect on knowledge creation and this link is mediated by knowledge storage capability. Darroch [2] concluded that knowledge acquisition, knowledge dissemination and responsiveness to knowledge have a positive impact on organizational innovation. Kianto [5] found a connection between knowledge management activities and continuous innovation.

The aim of this chapter is to handle the aforementioned challenges and queries whether innovation is significant on knowledge management. The survey study is designed to collect the required data from different companies located in different provinces of Iran.

MAIN BODY

Hypotheses

Here, we can create two hypotheses:

H1. Knowledge creation mediates the link between other knowledge processes and innovation.

H2. Each of four knowledge processes – knowledge creation, internal knowledge sharing, external knowledge acquisition and knowledge storage and documentation – has a direct impact on organizational innovativeness.

We can also illustrate these hypotheses in Figure 2.1.

We must mention that this modeling is not based on the standard method or popular modeling. These hypotheses also illustrate the relationship between innovation and knowledge management based on the literature. After that, just one concept still remains, and it is intensity. Based on the literature we can imagine two hypotheses about intensity:

H3a. The more knowledge intensive a company is, the more intense are its knowledge creation processes.

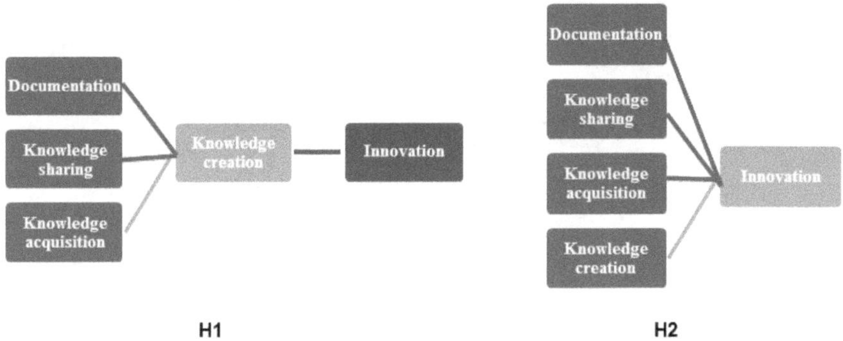

FIGURE 2.1 Research model for KM and innovation.

H3b. The more knowledge intensive a company is, the more intense are its knowledge sharing processes.
H3c. The more knowledge intensive a company is, the more intense are its knowledge acquisition processes.
H3d. The more knowledge intensive a company is, the more intense are its knowledge documentation processes.
H4a. The more knowledge intensive company is, the stronger is the impact of all knowledge processes towards innovation.

Every hypothesis is based on past works. These hypotheses refer to the concept that intensity has same effect on every component of the knowledge management model (H4a), or it has a weak relationship that cannot have the same power in every relation (H3a–H3d) [9]. A representation of the hypotheses is given in Figure 2.2.

In multiplying H1 and H2 with H3 and H4, we can infere the four separate models that are illustrated in Figure 2.3.

Analysis

To analyze these hypotheses one of the simplest methods is to use regression methods. Regression is a mathematical method that can find a quantity between every

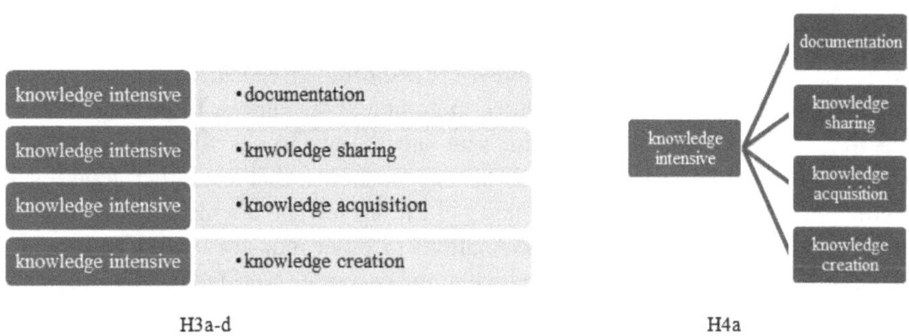

FIGURE 2.2 Research model for KM and intensity.

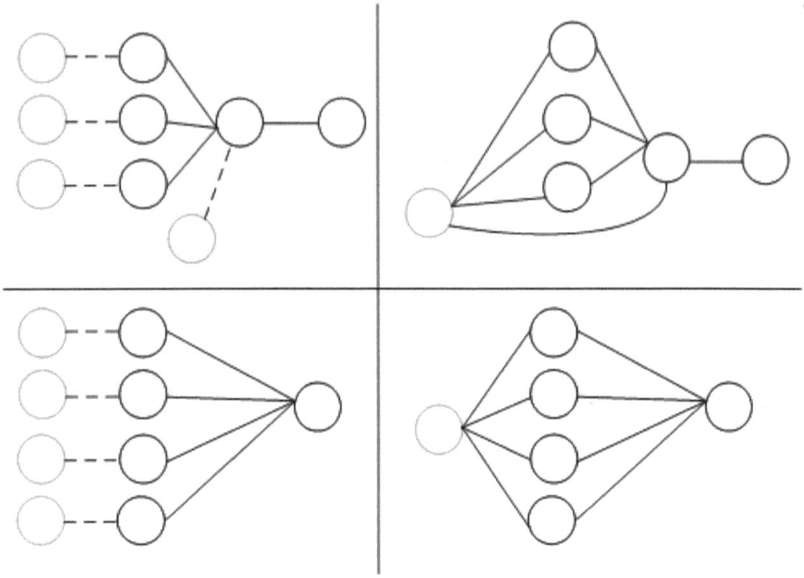

FIGURE 2.3 A four sub-model illustration.

cause and effect. In this research, one can assume every component as a cause and innovation as an effect. To complete these regression methods, one of the requirements is data collection. Data collections refer to points in hyper axes and the regression method can find coefficients with analyses on these points. The data was collected with a web-based survey in three provinces located in the north of Iran. Most of the past works focused on an individual population that can weaken the results. To obtain reliable, diverse and comparable data, it was decided to select companies with 50 or more employees that encompassed both the production and service sectors, and industries with different growth rates. As the first step, the pools of companies that fitted into the criteria described above were built based on publicly available databases. The size of the initial pool was 1,264 for province R and 10,000 in province M. These pools differed in size as a different response rate was expected across provinces. In province G, such a random pool was not used, due to the reasons described below. Next, the invitation letters explaining the purpose and the procedure of the research and providing the link to the web-based questionnaire were emailed to the selected companies. Respondents were promised an executive summary report of the research findings as an incentive to complete the survey. In province R, this was followed by two email reminders, sent one and two weeks after the initial mail. These resulted in 95 responses, or a 7.5% response rate; that is a rather good result, taking into account the significant length of the survey and absence of any informational support from any industry associations or other industry bodies. In province M, acknowledging the typical reluctance in the corporate world to participate in any research due to the culture of information secrecy, it was decided to have a bigger target random pool of companies. The software that was used for the administration of this survey allowed tracking the undelivered emails due to mistakes in the contact information or due to spam filters. It identified that out of 10,000 contacts selected from databases, only 4,064 had actually received the invitation email. This population yielded 145 visits to the survey page (3.6% of the population) and 21 responses (0.5% of the population or 14.5% of those who had visited the survey webpage). Taking into account the negative attitudes to this method of data collection in province M, multiplied by the length of the survey and the novelty of its subject area, this response rate, although very low, can be considered as good. Further on, to enlarge the province M sample, the invitation to participate in the survey was sent to the members of the alumni club of one of the province M business schools. This effort yielded a 0.6% response rate. Some respondents were also reached through the personal networks of the researchers (with a 66% response rate). As a result of these efforts, 83 responses were collected. In province G, similarly acknowledging the difficulty of the "cold call" research and the importance of personal networking, it was decided not to use random database mailing. The data collection was supported by a knowledge management center (KMC) in province G, the biggest online KM community in province G, which has about 1,000 members from different industries and regions. Additionally, some respondents were reached through the personal networks of the researchers. As a result of these efforts, 83 respondents from province G filled in this questionnaire. Taking into account properties of the data collection methods, the response rate via the online KM community can be estimated at 5%. As a result of data collection efforts, 261 responses in three provinces were collected. Of these, 40 responses were excluded from further analysis as they belonged to companies with under 50 employees. Therefore, the usable sample consisted of 221 responses, quite evenly representing the three provinces included in the survey (84 or 38% R, 64 or 29% M and 73 or 33% G responses). The organizations in the sample represent over 20 industries, with some domination of the production sector over the services sector (63% versus 37%). The majority of the companies employ between 50 and 200 employees (between 60 and 70% across three provinces). Around 70% of the companies in each of the three provinces are domestically owned. The survey reached the management level of the targeted organizations quite well: in provinces R and M over 70% of respondents belonged to middle or top management, and in province G – over 53%. The rest of the surveyed respondents, with minor exceptions, advised that they hold specialist positions in their organizations. While survey questions had been designed in a way that any employee of the organization could answer them, the high share of managerial responses made the data collected even more insightful. As the survey questions might have required some knowledge of the situation in the organization, the researchers controlled for the length of the respondent's service in the organization in the discussion. The majority of the respondents (93% in R and G, and 78% in M) had worked for their organization for more than one year. Therefore, the respondents from the sample provide a reliable picture of their organizations. Taking into account the diversity of the sample that consists of the responses from three very different provinces, where slightly different methods have been used to access the organizations, it was necessary to check for the potential differences among the subgroups in the sample. Differences between correlation and regression equations between the three provinces were examined, but no major differences were found [10].

Measurement

Every question for presenting the objective is a loan from previous works. Here, we have four components rather than innovation and intensity of knowledge management. A few items were extracted from [2] and [5], and the other ones were developed by the author. The scale for external knowledge acquisition was based on [5] and supported by the conceptual literature. It aimed to provide information on the frequency of knowledge-based interactions of the company with the external environment. In the first step, the correlation analysis should be performed. For this purpose, we used the Bartlett test of sphericity which demonstrated a highly significant number of correlations in the correlation matrix (see Tables 2.1 and 2.2).

In the second step for inferring hypothesis H2, we used Kaiser-Meyer-Oklin measure method. If H1 does not infer, H2 will infer. The results of regressions on H1 and H2 are illustrated in Figures 2.3 and 2.4. The hypothesized mediating effects (H1 and H2) were tested by using the mediated regression technique. The process for inferring of H1 is:

1. Regressing the mediator variable on the predictor variable (in this study documentation, knowledge sharing and knowledge acquisition on knowledge creation, respectively).
2. Regressing the criterion variable on the predictor variable (e.g., documentation on innovation).
3. Regressing the criterion variable simultaneously on the predictor and mediator variables (e.g., documentation and knowledge creation on innovation).

We should consider that H1 will infer if:

1. There is a significant relationship between the mediator and predictor variables (step 1).
2. There is a significant relationship between the predictor and criterion variables (step 2).
3. The mediator is significantly related to the criterion variable (step 3).
4. The effect of the predictor on the criterion variable is less in step 3 than in step 2. Full mediation occurs if the effect of the criterion variable is not significant in step 3. Partial mediation occurs if the criterion effect is reduced but significant.

Table 2.3 presents the mediated regressions demonstrating that while all knowledge processes impact innovation, knowledge creation fully mediates the impact of documentation, knowledge sharing and knowledge acquisition on innovation. When innovation is regressed simultaneously on the predictor and mediating variables (see Table 2.3, equation 1, step 3), the relationship between documentation and innovation decreased in magnitude (from B = 0.439 to B = 0.102), and becomes insignificant. The same finding applies to the relationship between knowledge sharing and innovation; the initially significant link between knowledge sharing and innovation (B = 0.596) becomes insignificant (B = 0.105) when knowledge creation is entered

TABLE 2.1
Factor Loadings and Coefficient Alphas of Knowledge Process Scales

	Component			
Items	**1**	**2**	**3**	**4**
Intra-organizational knowledge sharing and application (Cronbach α = 0.877)				
In our organization information and knowledge are actively shared within the units.	0.566	0.431	0.236	20.023
Different units of our organization actively share information and knowledge among each other.	0.602	0.373	0.366	20.016
In our organization employees and managers exchange a lot of information and knowledge.	0.687	0.335	0.179	20.014
Our organization shares a lot of knowledge and information with strategic partners.	0.614	0.275	0.149	0.290
Our employees are systematically informed of changes in procedures, instructions, and regulations.	0.690	0.230	0.353	0.145
Knowledge creation (Cronbach α = 0.868)				
Our organization frequently comes up with new ideas about our products and services.	0.121	0.786	0.242	0.170
Our organization frequently comes up with new ideas about our working methods and processes.	0.216	0.723	0.277	0.039
If a traditional method is not effective anymore, our organization develops a new method.	0.334	0.734	0.187	0.123
Our organization uses existing knowhow in a creative manner for new applications.	0.400	0.628	0.129	0.297
Knowledge storage and documentation (Cronbach α = 0.870)				
Our organization does a lot of work to refine, organize and store the knowledge collected.	0.388	0.351	0.644	0.126
Our organization possesses many useful patents and licenses.	0.036	0.170	0.696	0.171
In our organization, we are used to documenting in writing the things that are learned in practice.	0.145	0.248	0.832	0.203
In our organization, we make sure that the most important experiences gained are documented.	0.293	0.244	0.743	0.202
Knowledge acquisition (Cronbach α = 0.736)				
Our organization regularly captures knowledge of our competitors.	−0.022	0.355	0.116	0.701
Our organization regularly captures knowledge obtained from public research institutions including universities and government laboratories.	0.149	0.053	0.310	0.659
Our organization regularly captures knowledge obtained from other industry sources such as industrial associations, competitors, clients and suppliers.	0.264	0.135	0.228	0.647

Note: (n = 221)

TABLE 2.2
Means, Standard Deviations and Correlations between Variables

	Mean	SD	1	2	3	4	5
Innovation performance	3.60	1.08	1.00				
Knowledge creation	3.94	1.16	0.664***	1.00			
Documentation and storage	3.54	1.30	0.408***	0.575***	1.00		
Knowledge sharing	3.87	1.33	0.513***	0.689***	0.623***	1.00	
Knowledge acquisition	3.70	1.18	0.313***	0.485***	0.495***	0.461***	1.00
Knowledge intensity	4.41	1.11	0.277***	0.407***	0.202**	0.356***	0.158*

Notes: *p < 0.05; **p < 0.01; ***p < 0.001

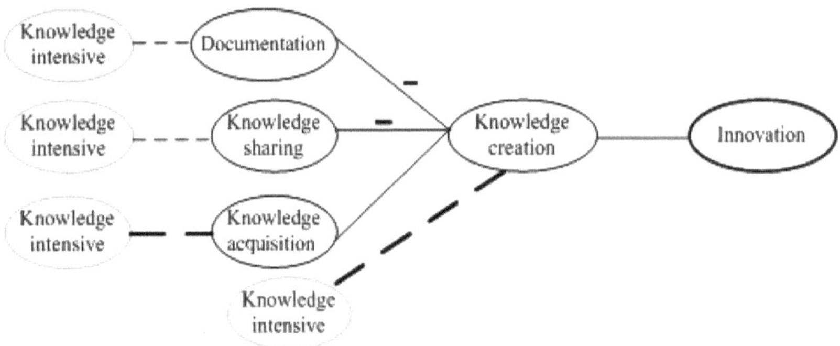

FIGURE 2.4 The finalized research model.

into the equation (equation 2, step 3). Similarly, the relationship between knowledge acquisition and innovation turns insignificant in step 3 (from B = 0.345 to B = 0.013). Based on these results, H1is accepted and H2 is rejected.

H3a–H3d predicted that knowledge intensity impacts all knowledge processes. This assertion was examined by regressing each knowledge process onto knowledge intensity. The results, presented in Table 2.4, demonstrate that knowledge intensity positively impacts documentation, knowledge sharing, knowledge acquisition and knowledge creation. While the impact on documentation and knowledge acquisition is rather small, knowledge intensity explains 2 to 4% of their variation which is statistically significant. On the other hand, knowledge intensity seems to have quite a large impact on knowledge sharing and knowledge creation (12 and 16% respectively). Thus, H3a–H3d are accepted.

Support for a moderation hypothesis would exist when (see Table 2.5):

TABLE 2.3
Mediating Effects of Knowledge Creation on Innovation

Step	Criterion	Predictor	B	T	Adj.R^2
1	Knowledge creation	Documentation.	0.619	11.589*	0.381
2	Innovation	Documentation	0.439	7.188*	0.189
3	Innovation	Documentation	0.102	1.768	0.510
		Knowledge creation	0.654	11.307*	
1	Knowledge creation	Knowledge sharing	0.723	15.373*	0.520
2	Innovation	Knowledge sharing	0.596	10.857*	0.352
3	Innovation	Knowledge sharing	0.105	1.503	0.444
		Knowledge creation	0.594	8.539*	
1	Knowledge creation	Knowledge acquisition	0.543	9.508*	0.292
2	Innovation	Knowledge acquisition	0.345	5.404*	0.115
3	Innovation	Knowledge acquisition	0.013	0.239	0.498
		Knowledge creation	0.703	12.771*	

Notes: *$p < 0.001$

TABLE 2.4
Regression Results for Knowledge Intensity and Knowledge Processes

Criterion	Predictor	B	T	Adj.R^2
Documentation	Knowledge intensity	0.202	3.031**	0.036
Knowledge sharing	Knowledge intensity	0.356	5.602***	0.123
Knowledge acquisition	Knowledge intensity	0.158	2.358*	0.021
Knowledge creation	Knowledge intensity	0.407	6.554***	0.161

Notes: *$p < 0.05$; **$p < 0.01$; ***$p < 0.001$

- The results of the model are significant;
- The interaction term is significant in the hypothesized direction; and
- The values for the changes in R^2.

CONCLUDING REMARKS

In this chapter, we discussed the research on knowledge management and analyzed its effect on innovation and intensity. After a review of the literature, four different hypotheses were inferred in two areas and resulted in four different models. Then, using the regression method, we inferred that H2 and H3a–H3d are accepted. The finalized model is depicted in Figure 2.4.

TABLE 2.5
Results of the Moderator Regression Analyses

Model	Criterion	Predictor	B	T	Adj.R^2	Change R^2
1	Knowledge creation	Documentation	0.659	13.499***	0.431	
2	Knowledge creation	Documentation	0.626	13.058***	0.475	0.046***
		Knowledge intensity	0.217	4.531***		
3	Knowledge creation	Documentation	0.994	7.480***	0.492	0.019**
		Knowledge intensity	0.570	4.444***		
		Knowledge intensity * Documentation	–0.563	–2.957**		
1	Knowledge creation	Knowledge sharing	0.747	17.303***	0.556	
2	Knowledge creation	Knowledge sharing	0.747	15.994***	0.568	0.014**
		Knowledge intensity	0.122	2.723**		
3	Knowledge creation	Knowledge sharing	1.037	9.328	0.584	0.018**
		Knowledge intensity	0.496	3.918**		
		Knowledge intensity * Documentation	–0.570	–3.151**		
1	Knowledge creation	Knowledge acquisition	0.629	12.446***	0.393	
2	Knowledge creation	Knowledge acquisition	0.592	11.728***	0.429	0.039***
		Knowledge intensity	0.201	3.982***		
3	Knowledge creation	Knowledge acquisition	0.694	4.765***	0.428	0.001
		Knowledge intensity	0.298	2.149*		
		Knowledge intensity * Documentation	–0.158	–0.749		
1	Innovation performance	Knowledge creation	0.667	13.797***	0.445	
2	Innovation performance	Knowledge creation	0.669	12.955***	0.441	0.000
		Knowledge intensity	–0.005	–0.099		
3	Innovation performance	Knowledge creation	0.752	6.042***	0.440	0.000
		Knowledge intensity	0.094	0.651*		
		Knowledge intensity * Documentation	–0.152	–0.735		

Notes: *$p < 0.05$; **$p < 0.01$; ***$p < 0.001$

The somewhat surprising finding can be interpreted regarding exploitative and explorative knowledge application processes in organizations. Perhaps firms in less knowledge-intensive conditions have to exploit more the explicit (documented) and tacit (made collective by knowledge sharing) knowledge which already exists in their company for knowledge-creation purposes. While firms in highly knowledge-intensive conditions should rather explore new knowledge as material for knowledge creation, and therefore do not have so much use for existing knowledge in the firm.

REFERENCES

1. Trott P. *Innovation management and new product development.* 3rd ed. New York, NY: Pearson Education Limited; 2005.
2. Darroch J. Knowledge management, innovation and firm performance. *J Knowl Manag.* 2005;9(3):101–15.
3. Basadur M, Gelade GA. The role of knowledge management in the innovation process. *Creat Innov Manag.* 2006;15(1):45–62.
4. Marqués DP, Simón FJG. The effect of knowledge management practices on firm performance. *J Knowl Manag.* 2006;10(3):143–56.
5. Kianto A. The influence of knowledge management on continuous innovation. *Int J Tech Manag.* 2011;52(1/2):7–13.
6. Chou S-W. Knowledge creation: absorptive capacity, organizational mechanisms, and knowledge storage/retrieval capabilities. *J Inf Sci.* 2005;31(6):453–65.
7. Qianwang D, Dejie Y. An approach to integrating knowledge management into the product development process. *J Knowl Manag Prac.* 2006;7(2):26–34.
8. Deng X, Doll WJ, Cao M. Exploring the absorptive capacity to innovation/productivity link for individual engineers engaged in IT enabled work. *Inf Manag.* 2008;45(2):75–87.
9. Taminiau Y, Smit W, de Lange A. Innovation in management consulting firms through informal knowledge sharing. *J Knowl Manag.* 2009;13(1):42–55.
10. Fazlollahtabar H, Aghasi E. *Knowledge management: from concept to engineering implementation.* Mashhad, Iran: Sabet Ghadam Publishing; 2012. ISBN: 978-600-91153-8-9 (in Persian).

3 Knowledge with Innovation Performance

INTRODUCTION

Due to the rapid changes in the business environment, the capability to achieve innovation quickly through external knowledge has become a key determinant of competitive advantage in the era of the knowledge economy. Previous research has identified some factors that contribute to a firm's innovation performance, including relationship learning and effective communication among different companies engaging in learning activities in the context of technological changes.

The recognition of the significance of external knowledge in a firm's outside network is an important phenomenon seen in the organization of the innovation process since the 1980s [1]. External knowledge is a very important resource for learning new techniques, solving problems, creating core capabilities and initiating new situations for organizations. To turn knowledge into actions and have better innovation performance, organizations may need to understand the main characteristics of their knowledge.

Several previous studies have considered that a firm's innovation performance depends not on the knowledge that the firm may accumulate, but on its abilities to turn that knowledge into action [2–4]. There was no previous study exploring the influences of the properties of knowledge and the absorptive capacity upon innovation performance [5]. The focus of this research is to find the critical role of absorptive capacity in determining a firm's innovation performance. Hence, the two antecedents of the research framework in this study are properties of knowledge and absorptive capacity, and the consequent is a firm's innovation performance. This study addressed three research questions:

First, are properties of knowledge negatively associated with a firm's innovation performance?

Second, is absorptive capacity positively associated with a firm's innovation performance?

Third, is absorptive capacity the moderator between the properties of knowledge and the consequent, a firm's innovation performance?

In the remainder of the chapter, we first explore in more detail the concepts of properties of knowledge and absorptive capacity, and how they expect them to be related to a firm's innovation performance. The author then outlines the methodology used in this study, followed by an examination of the results and data analysis. Finally, the authors conclude with a discussion of the implications of this study.

MAIN BODY

In the age of the knowledge economy, the most important asset of an organization is knowledge. How effectively and efficiently knowledge is managed to improve the innovation performance of an organization is the major issue of concern for firms in the twenty-first century. Most theorists agree that a firm's innovation performance depends not on the knowledge that the firm may accumulate but on its abilities to turn that knowledge into action [2–4]. To turn knowledge into actions, organizations may need to understand the main characteristics of their knowledge.

Properties of Knowledge

Researchers categorize knowledge into two types: explicit knowledge and implicit (tacit) knowledge. Compared to explicit knowledge, tacit knowledge, embedded in complex interactions, processes and routines within the firm, is rendered ambiguous and consequently creates barriers to imitation. Thus, it can only be sensed, observed and experienced [6]. The R&D staff who conduct a laboratory experiment may get much closer to documenting the process, but the codification is incomplete in the sense that the personal knowledge of the staff cannot be fully included in an explicit message [7].

Tacit knowledge and the characteristics of the parties involved in the transfer of knowledge have been extensively studied. For example, researchers delineated that the tacit characteristic of knowledge determines the costs and mode of sharing. The conclusion of the research suggested that the sharing of manufacturing capabilities is influenced by the degree to which they may be codified. Another research also found that tacit knowledge will not be easier to communicate. Furthermore, they suggested that transfer of tacit knowledge may require a robust and iterative bidirectional communication between the parties. They stated that further research is necessary to study other properties of knowledge (like its ambiguity or its complexity) and find the impacts that these characteristics will have in the transfer of information [8].

Causal ambiguity, as an important property of knowledge, had also been studied. Yet, as eloquently stated by Barney, causal ambiguity has been a concept in the strategic management and organization theory literature for a period of time. However, the full implications of this concept have been largely undeveloped. In the literature, researchers explored the fundamental characteristics of knowledge development. When knowledge is communicated and it has a higher causal ambiguity, its interchange is more problematic because of the difficulty in understanding the whole concept [9].

Basically, as with causal ambiguity, it encapsulates a similar lack of understanding of the logical linkages between actions and outcomes, inputs and outputs and causes and effects that are related to R&D. The higher degree of ambiguity may also reduce the propensity to learn from a partner. That is, when the degree of ambiguity associated with a partner's knowledge is high, the chances of effectively repatriating and absorbing the competence are rather limited. So, ambiguity is negatively related to knowledge transfer and a firm's innovation performance.

Another important property of knowledge is complexity. Knowledge complexity is defined as the number of interdependent technologies, routines, individuals and resources linked to a particular knowledge or asset. Competence may span numerous individuals and departments so that the totality of the knowledge cannot be easily integrated or understood by many individuals. As knowledge becomes more complex, organizations need to absorb more areas of knowledge content, as well as understand the interlinkages between the different content areas. Thus, complexity is expected to affect the comprehension of the totality of an innovation process and to impair its transferability. Simple knowledge is easier to absorb than complex knowledge [10].

According to the above researchers, the authors use three dimensions (tacitness, ambiguity and complexity) to characterize properties of knowledge and based on these dimensions develop questionnaire items. The difficulties of using high a degree of tacit, ambiguous and complex knowledge are illustrated by the numerous failures of R&D projects. For instance, it seems to be quite common that such R&D projects aiming at making vast bodies of tacit knowledge explicit run into serious difficulties. Without this transformation process, many R&D projects fail. Examples can be found both in the information technology industry and electro-communication industry. Consistent with arguments based on the properties of knowledge, the authors expect that common knowledge (with the lower degree of tacitness, ambiguity and complexity) will ease the transfer of knowledge. Hence:

H1. The degree of properties of knowledge (tacitness, ambiguity and complexity) is negatively related to firm innovation performance.

Absorptive Capacity

This study is concerned specifically with how absorptive capacity is related to innovation performance. The researchers argued that absorptive capacity is critical to a firm's innovation process. Specifically, absorptive capacity is defined as "the ability of a firm to recognize the value of new, external information, assimilate it and apply it to commercial ends". Given that, external knowledge is an important resource for firm innovation.

Absorptive capacity increases the speed and frequency of incremental innovation because such innovations draw primarily on the firms' existing knowledge base. Given that absorptive capacity has also been operationalized regarding knowledge content (i.e., patents), it is not surprising that several studies have shown significant support for the hypothesis that absorptive capacity positively affects innovation.

The researchers defined absorptive capacity as a set of organizational routines by which firms acquire, assimilate, transform and exploit external knowledge from outside networks to produce a dynamic organizational capacity. Through absorptive capacity, firms can learn from inter-organizational partners and expand their knowledge and skill, improve their ability to assimilate, to utilize future external knowledge and information and eventually to enhance their innovation performances. Firms exposed to the same amount of external knowledge might not derive equal benefits, because they differ in their ability to identify and exploit such knowledge.

In other words, absorptive capacity can be a source of a firm's better innovation performance. The authors posit that absorptive capacity has an impact on innovation performance only when there is external knowledge from outside networks that can be acquired, assimilated, transformed, and after that, exploited. Firms with greater absorptive capacity will benefit more from the use of tacit, ambiguous and complex external knowledge. Based on the above discussion concerning absorptive capacity and the empirical evidence from prior literature, the authors would expect that absorptive capacity has a moderating effect between knowledge's properties and innovation performance. Therefore, this study proposed the following hypothesis:

H2. The relationship between the properties of knowledge and innovation performance is more pronounced when the firm has a higher absorptive capacity.

The Research Framework of the Study

This study proposed two hypotheses and showed the research framework in Figure 3.1. There was no previous research exploring the effects of the external factor (properties of knowledge) and the internal factor (absorptive capacity) on innovation performance. This study discussed properties of the effects knowledge on the innovation performance through the moderator, absorptive capacity, in Iranian small and medium-sized enterprises (SMEs).

METHODS AND MEASUREMENT

Data Collection and the Sample

To test the predictions, the authors collected data from SMEs in Iran during 2010–2017. The sampling frame consisted of all enterprises operating in two industries: information technology and electro-communication. The authors selected these industries because they were similar regarding their composition (e.g., number of employees, average age, customer base).

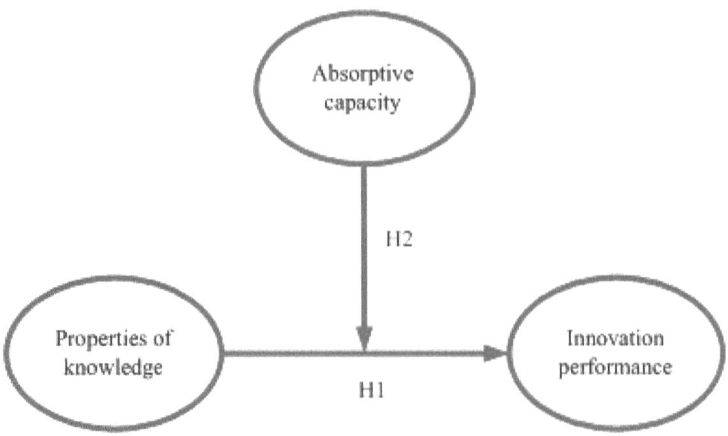

FIGURE 3.1 The research framework.

The measurement of the questionnaire items in this study was by use of the "seven-point Likert scale from 1 to 7", rating from "strongly disagree" to "strongly agree". A total of 206 people responded to the request for information about their company. This number is approximately 32% of the original 638 respondents. Complete responses were obtained from 96 firms. All of the firms from which the authors obtained responses have 200 or fewer employees. Further, 47% of the firms had 100 or fewer employees. On average, firms in the sample had sales of US$50,000 and had been in operation for ten years. Over 43% of the firms in the sample were privately owned.

Definitions and Measurements of the Constructs

This study uses a multi-measure approach to operationalize the theoretical constructs. Whenever possible, measurement instruments available from extant research are used to operationalize the theoretical constructs. The questionnaire comprises four parts. The first part of the questionnaire consists of the descriptive data of companies; the second part is the measurement of properties of knowledge; the third part is the measurement of absorptive capacity; the fourth part is the innovation performance. The definitions and measurements of the constructs are further defined as follows.

Dependent Variables

The authors use two proxies aimed at reflecting various types of innovative performance by firms. First, the authors use a variable aimed at indicating the ability of the firm to produce technical innovations. This variable is measured as the firm's new products.

Independent Variables

The constructs of tacitness, ambiguity and complexity are chosen, as they are the aspects that can be most expected to influence the firm's innovation performance.

Moderate Variables

Moderate variables in this study are related to firm's absorptive capacity, which is what enables the companies to effectively acquire and utilize external as well as internal knowledge, which affects the company's ability to innovate [5]. This study defines absorptive capacity as "the ability to acquire, to assimilate, to transform and to exploit knowledge" that may determine its levels of organizational innovation [5]. The authors view absorptive capacity as being acquired through a multistage, organizational process. To develop a variable for statistical analysis, the author creates a seven-point scale that captures three primary stages associated with the absorptive capacities: acquisition, assimilation, transformation and exploitation. First, the author uses the acquisition to reflect a firm's capability of identifying and acquiring externally generated knowledge that is critical to its operations. Second, assimilation is used to capture the features of a firm's routines and processes that allow it to analyze the process, interpret and understand the information obtained from external sources. Finally, the author takes transformation and exploitation to denote a firm's capability to develop and refine the routines that use combining existing knowledge and the newly acquired and assimilated knowledge to create new competencies.

RESULTS

Model and Analysis

The basic variables in this chapter are collected from the same source. To assess common-variance (CMV) among these variables, the author conducts Harman's post hoc single factor test. A principal component exploratory factor analysis shows three factors that have eigenvalues greater than one and explained 69.12% of the variance, with the largest factor explaining 31.14%. The items loaded appropriately as the author expected and the factor loadings are generally high with the lowest being 0.55. The measures show good internal reliability, with alpha coefficients ranging from 0.715 to 0.892.

This study utilized multiple regression (MR) and moderated multiple regression (MMR) analysis to verify the research framework and hypotheses. The antecedent of the research framework in this study is properties of knowledge, and the consequent is innovation performance, while the moderator is absorptive capacity. The model is hypothesized and estimates of the parameter values are used to develop an estimated regression equation, between a dependent variable (innovation performance) and independent variables (properties of knowledge). The F-test gives values for the independent variables when it is satisfactory; it means that the estimated regression equation can be used to predict the value of the dependent variable. Then, to examine the influence of the main effects of variables and interaction variables, separately, the author uses MMR in three steps. The first step: the author models the firm's innovation performance as a function of independent variables (M1 – enter with the independent variables in step 1). Second, they model the firm's innovation performance as a function of independent and moderate variables (M2 – enter with the independent variables and moderator in step 2). The third step: they introduce interaction effects – innovation performance with absorptive capacity (moderate effects) – to the main effects (M3 – enter with the interactions between independent variables and moderator in step 3).

Table 3.1 presents the details of the measurement instruments and scales used to operationalize the theoretical constructs. The Cronbach's α reliabilities for each construct are also reported in Table 3.1. Except for the tacitness and ambiguity construct, which were 0.715 and 0.772, constructs were at or above the value of 0.801. Because the Cronbach's α coefficients of all constructs were more than 0.7, the measurement of this study was acceptable in the reliability aspect. On the other hand, it is also important to verify the validity of the measurement.

Table 3.2 shows the correlation matrix. There were significantly positive correlations among properties of knowledge, absorptive capacity and innovation performance. The correlation analysis performs the significant relationship between the two variables, and it cannot mean the variables have a causal relationship. In the next section, multiple regression analysis will be applied for the causal relationships test between the variables.

TABLE 3.1
Measurement Instruments

Variables	Measurement Items	Internal Consistency Reliability (α)
Properties of knowledge	Tacitness	0.715
	Ambiguity	0.772
	Complexity	0.807
Absorptive capacity	Acquisition	0.834
	Assimilation	0.874
	Transformation and exploitation	0.801
Innovation performance		0.892

Regression Analysis of Innovation Performance on Properties of Knowledge

First, the results that the author tested for the effects of innovation performance for analyzing properties of knowledge are presented. The adjusted R^2 of the first model is 0.144. There are statistically significant and negative effects of technology innovation performance for the analysis of tacitness, but there are significant and positive effects of technology innovation performance for analyzing ambiguity and complexity. Contrary to the previous hypothesis, when the knowledge is more ambiguous and complex, it can produce better technology innovation performance. The data is shown in Table 3.3.

Then, from Table 3.4, the adjusted R^2 of this model is 0.133. There are statistically significant and negative effects of management innovation performance for the analysis of ambiguity, and there are no statistically significant management innovation performance effects for the analysis of tacitness. Contrary to the previous hypothesis, again, Table 3.4 exhibits the effects of complexity on management innovation performance that affect technology innovation performance as well. Complexity has significant and positive effects on management innovation performance.

The Moderate Effects of a Firm's Absorptive Capacity

In this section, we present the results of testing the moderate effects of a firm's absorptive capacity between the properties of knowledge and innovation performance. ZKT × ZCAT is the moderate effect of a firm's acquisition capacity between knowledge tacitness and innovation performance; ZKA × ZCAT is the moderate effect of a firm's acquisition capacity between knowledge ambiguity and innovation performance; ZKC × ZCAT is the moderate effect of a firm's acquisition capacity between knowledge complexity and innovation performance (Table 3.5). There is a significantly interactive effect between knowledge tacitness and moderators of

TABLE 3.2
Correlations and Constructs

Constructs	1	2	3	4	5	6	7	8
Tacitness	1							
Ambiguity	0.616**	1						
Complexity	0.551**	0.579**	1					
Acquisition	0.237**	0.126	0.321**	1				
Assimilation	0.296**	0.215**	0.393**	0.734**	1			
Transformation and exploitation	0.213**	0.092	0.333**	0.577**	0.764**	1		
Technical innovation performance	0.208*	0.15	0.322**	0.697**	0.711**	0.642**	1	
Management innovation performance	0.201*	0.096	0.310**	0.538**	0.680**	0.651**	0.675**	1

Notes: $*p < 0.05$; $**p < 0.01$

TABLE 3.3
Regression Analysis of Technology Performance on Properties of Knowledge

Properties of Knowledge		Beta Coefficient	Sig.	VIF
Tacitness		–0.160	0.003*	1.852
Ambiguity		0.266	0.001**	1.959
Complexity		0.293	0.002*	1.717
R^2	0.195			
Adjusted R^2	0.144			
Sig.	0.000**			
F-value	3.795			

Notes: *$p < 0.01$; **$p < 0.001$

TABLE 3.4
Regression Analysis of Management Innovation Performance on Properties of Knowledge

Properties of Knowledge		Beta Coefficient	Sig.	VIF
Tacitness		0.098	0.342	1.831
Ambiguity		-0.179	0.041*	1.925
Complexity		0.338	0.001**	1.667
R^2	0.168			
Adjusted R^2	0.133			
Sig.	0.000**			
F-value	4.833			

Notes: *$p<0.05$; **$p<0.001$

acquisition capacity on innovation performance. The higher the level of acquisition capacity and the lower level of tacit knowledge, the better the performance of innovation accomplishment, but there is no significantly interactive effect between another two properties of knowledge and moderators of acquisition capacity on innovation performance.

CONCLUDING REMARKS

The empirical results of this study showed that little support was received for H1. Only a few of the properties of knowledge have negative effects on innovation performance, but most of the properties of knowledge have a positive effect on competitive advantage. Hence, these results indicated that the higher the degree of the ambiguous and complex knowledge, the better is the technology innovation performance.

TABLE 3.5
The Moderate Effects of Firm's Assimilation Capacity

Testing Sequence	M_1	M_2	M_3
[a]ZKT			
Beta	0.224***		–0.002
ΔR^2	0.044		0.572
Sig.	0.006		0.968
ZCAT			
Beta		0.762***	0.763***
ΔR^2		0.575	0.572
Sig.		0.000	0.000
ZKT × ZCAT			
Beta			–0.204*
ΔR^2			0.572
Sig.			0.050
ZKA			
Beta	0.146*	–0.019	–0.009
ΔR^2	0.015	0.575	0.585
Sig.	0.073	0.731	0.868
ZCAT			
Beta		0.766***	0.727***
ΔR^2		0.575	0.585
Sig.		0.000	0.000
ZKA × ZCAT			
Beta			–0.704**
ΔR^2			0.585
Sig.			0.037
ZKC			
Beta	0.346***	0.055	0.060
ΔR^2	0.120	0.577	0.577
Sig.	0.000	0.345	0.304
ZCAT			
Beta		0.740***	0.736***
ΔR^2		0.577	0.577
Sig.		0.000	0.000
ZKC × ZCAT			
Beta			–0.060
ΔR^2			0.577
Sig.			0.349

Notes: *$p < 0.1$; **$p < 0.05$; ***$p < 0.01$

[a] ZKT = Standardized value of tacitness; ZKA = Standardized value of ambiguity; ZKC = Standardized value of complexity; ZCAT = Standardized value of assimilation.

The marvelous capacity of the human mind to make sense of a lifetime's collection of the higher degree of ambiguous and complex knowledge is, by its very nature, hard to capture. However, it is essential to the technology innovation process. At every stage, technology innovation requires a solution, convergence upon acceptable action, ambiguous and complex knowledge plays an important role. The director of an advanced R&D group commented that his researchers were likely to be "stuck for life" with a technology they created because the knowledge base moves so fast it is never totally captured in any simple form.

Where the process of eliciting and managing the flow of the higher degree of tacit, ambiguous and complex knowledge is easy, the innovation would still not occur effortlessly – but it would be much less of a challenge. Multiple barriers exist both to the stimulation of divergent thinking and then to the coalescence of that thinking around a common aim. Previous studies have considered that a firm's innovation performance depends not on the knowledge that the firm may accumulate but on its abilities to turn that knowledge into action.

The author would, therefore, expect that absorptive capacity, which is the firm's ability to acquire, assimilate and profitably utilize a higher degree of the ambiguous and complex knowledge, appears to be one of the most important determinants of increasing innovation performance.

This chapter looks at the moderate role played by absorptive capacity, as an avenue to measure its impact on innovation performance through properties of knowledge. The MMR presented some critical roles played by absorptive capacity as a part moderator of properties of knowledge on innovation performance. The author found that different kinds of absorptive capacity have different moderating effects on knowledge and innovation performance. First of all, the higher level of acquisition and assimilation capacity and the lower level of tacit knowledge bring better performance of innovation accomplishment (20.204). Second, the higher level of assimilation capacity and the lower level of ambiguous knowledge brings better performance of innovation accomplishment. And third, the higher level of transformation and exploitation capacity and the lower level of ambiguous and complex knowledge, the better the performance of innovation accomplishment. These empirical results show that firms endowed with more acquisition and assimilation capacity are better equipped to identify the presence of external tacit knowledge. These results also show that firms which have more assimilation capacity are better equipped to absorb ambiguous external knowledge. Finally, this research finds that firms that possess more transformation and exploitation capacity are better equipped to utilize the external ambiguous and complex knowledge. Therefore, with few exceptions, H2 is supported.

Beyond theoretical and empirical contributions to these two kinds of literature, this work has practical importance for managers. This study provides new insights to managers, that by promoting their firms' special absorptive capacity to the particular properties of knowledge, their innovation performance will be increased. The practical implications of the study are expected to be most evident for information technology and electro-communication industries in which technology is moving quickly and collaborative innovation has been and continues to be a significant feature. Consequently, building appropriate kinds of absorptive capacity is a tough job for most of the organizations. For most of the tacit, ambiguous and complex knowledge that is difficult for internal staff to assimilate, a gatekeeper both monitors the outside networks and translates the knowledge into a form understandable to the research group.

However, when external knowledge flows are somewhat random, and it is not clear where in the firm or subunit a piece of outside knowledge is best applied, a centralized gatekeeper may not provide an effective link to the outside networks. Moreover, even when a gatekeeper is important, his or her absorptive capacity does not constitute the absorptive capacity of his or her unit within the firm and cannot fully understand the tacit, ambiguous and complex knowledge individually. Therefore, relying on a small set of gatekeepers may not be sufficient; the group as a whole must have some level of relevant background knowledge, and when knowledge properties are highly tacit, ambiguous and complex, the requisite level of background may be rather high. It has become generally accepted that complementary functions within the firm ought to be tightly intermeshed, recognizing that some amount of redundancy in expertise may be desirable to create what can be called cross-function absorptive capacities. For example, close linkages between R&D and manufacturing are often helpful the exploiting ambiguous and complex knowledge from the firm's outside network and credit for the relative success of firms in moving products rapidly from the design stage through development and manufacturing.

Finally, to integrate certain classes of tacit, ambiguous and complex knowledge successfully into the firm's activities, the firm requires an existing internal staff of technologists, scientists and managers who are both competent in their fields and are familiar with the firm's idiosyncratic needs, organizational procedures, routines, complementary capabilities and extramural network relationships.

This research has important practical implications for Iranian managers. The results of the study are consistent with the observation that Iran is moving from a person-based to a network-based society. In this network-based economy, where a large proportion of relevant knowledge resides outside a firm's boundaries, this is a particularly important message for Iranian managers who aim to develop a sustainable competitive advantage. Managers must recognize the important value of complex knowledge in the firm's outside network in their managerial innovation. External knowledge is a very important resource for learning new techniques, solving problems, and creating core capabilities in the Iranian new business environment. Meanwhile, to turn knowledge into actions and have better innovation performance, organizations may need to understand the main characteristics of their knowledge and figure out the relationship with different kinds of absorptive capacity. The author's findings also highlight that absorptive capacity will be relatively more important in turbulent knowledge environments when intellectual property rights (IPR) protection is stronger in Iran. IPR protection makes the outside knowledge more tacit, ambiguous and complex than before. Iranian governments set on formulating policies to foster firms' absorptive capacity would be well advised to do so in high-tech sectors (especially in information technology and electro-communication sectors), in conjunction with initiatives aimed at increasing IPR protection. High-tech SMEs in Iran are required to seek more cooperation with other partners, such as research institutions, universities and intermediary institutions by establishing cooperation networks.

There are multiple avenues for further research in this area. First, this chapter does not address the effects of environmental conditions on the absorptive capacity and innovation performance relationship. Environmental factors, such as the outside network topology, will determine the access to new knowledge. A well-established type of network topology can promote the transfer of external knowledge. By occupying a central network position, a corporation is likely to access useful knowledge from other network members. Second, the author needs to do further work to find out whether the results in this chapter are different for companies in the same industry in another country. Moreover, further research might also explore the importance of the network structure of the linkages between university scientists and firm researchers. Theory on strategic networks suggests that linkages to network partners may increase the breadth and variety of information to which a firm has access, while strong linkages to one or a few network partners may unproductively limit a firm's vision of alternatives. Finally, research using inter-organizational networks is still constrained by the limitations of considering ties at a single level of analysis – the firm. The cross-level effects between networks at different levels of analysis are a fertile area of study that is only beginning to be meaningfully addressed in the field.

REFERENCES

1. Escribano A, Fabrizio KR. Absorptive capacity and the search for innovation. *Res Pol.* 2009;38(2):255–67.
2. Alavi M, Leidner D. Review: knowledge management and knowledge management systems: conceptual foundations and research issues. *MIS Quarterly* 2001;25(1):107–36.
3. Argote L, McEvily B, Reagans R. Managing knowledge in organizations: an integrative framework and review of emerging themes. *Manag Sci.* 2003;49(4):571–82.
4. Ari J. Knowledge-processing capabilities and innovative performance: an empirical study. *Eur J Innov Manag.* 2005;8(3):336–49.
5. Daghfous A. Absorptive capacity and the implementation of knowledge-intensive best practices. *SAM Adv Manag J.* 2004;69(2):21–7.
6. Ahuja G, Katila R. Technological acquisitions and the innovation performance of acquiring firms: a longitudinal study. *Strat Manag J.* 2001;22(3):197–220.
7. Gans J., Stern S. The product market and the market for 'ideas': commercialization strategies for technology entrepreneurs. *Res Pol.* 2003;32(2):333–50.
8. Hansen MT. The search-transfer problem: the role of weak ties in sharing knowledge across organization subunits. *Admin Sci Q.* 1999;44(1):82–111.
9. Dyer JH, Singh H. The relational view: cooperative strategy and sources of inter-organizational competitive advantage. *Acad Manag Rev.* 1998;23(4):660–79.
10. Fazlollahtabar H, Aghasi E. *Knowledge management: from concept to engineering implementation.* Mashhad, Iran: Sabet Ghadam Publishing; 2012. ISBN: 978-600-91153-8-9 (in Persian).

4 Knowledge and Innovation in Networked Environments

LITERATURE REVIEW

From Dynamic Capabilities to Knowledge-Based Dynamic Capabilities

In [1] a pioneering researcher of this field defined dynamic capabilities as "the ability to integrate, build and reconfigure internal and external capabilities to address rapidly changing environments". According to [1] and [2] by this definition, the object of dynamic capabilities is not clear and includes a kind of resource in the general sense. Also, the performance consequence of dynamic capabilities was uncertain. Researchers in [3] defined dynamic capability as "a learned and stable pattern of collective activity" to modify their operational processes and improve their effectiveness. Teece [4,5] modified his definition of dynamic capabilities as "the capabilities that enable business enterprises to create, deploy and protect the intangible assets that support superior and long business performance". The framework of dynamic capabilities overlaps with absorptive capacities that refer to the "ability to identify, assimilate and exploit knowledge from the environments". Substantial extensions were made, such as relative absorptive capacity, potential absorptive capacity and realized absorptive capacity [6].

The Construct of Knowledge-Based Dynamic Capabilities

As dynamic capability is the ability to acquire, generate and combine resources, three sub-capabilities – knowledge acquisition capabilities (KAC), knowledge generation capabilities (KGC), and knowledge combination capabilities (KCC) represent three dimensions of knowledge-based dynamic capabilities (KDC), building upon each other to produce the integrated dynamic capabilities of a firm (illustrated in Figure 4.1). Knowledge is the principal productive resource of the firm. Considering a firm's boundary, knowledge can be categorized into internal accumulated/generated knowledge and external knowledge. In fact, organizational knowledge exists in two different forms: explicit knowledge and tacit knowledge. The second component of KDC is the knowledge-generating capabilities in a framework. These capabilities differentiate one organization from the others, denoting a firm's ability to develop and refine the activities and processes that facilitate creating/generating new knowledge. The underlying processes include internal R&D, the socialization, externalization, combination and internalization (SECI) process proposed by Nonaka [7] and knowledge creation through external venturing [8]. The third part of KDC is

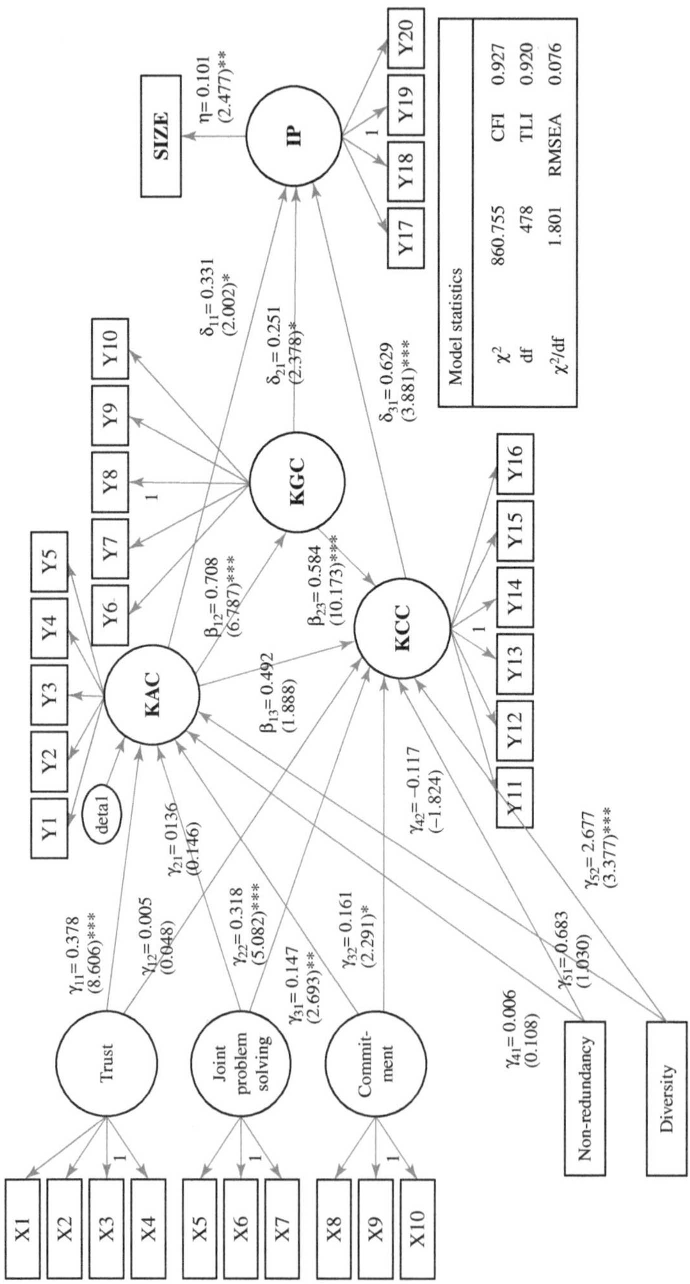

Note: $^{*}p < 0.05$; $^{**}p < 0.01$; $^{***}p < 0.001$, t-value in parentheses

FIGURE 4.1 A research model.

knowledge combination capability. Combination capability is the firm's ability to integrate and apply internal and external knowledge, sometimes resulting in producing a completely new knowledge.

Network Embeddedness and Innovation

In changing environments, innovation becomes the main part of the field of strategic management as its vital role in gaining and maintaining competitive advantage. Enhancing innovation capability and improving innovation performance is the focus of several researchers [9,10].

HYPOTHESES DEVELOPMENT

The Link between Knowledge-Based Dynamic Capabilities and Innovation

As mentioned previously, this chapter conceptualizes dynamic capabilities as a series of knowledge-based capabilities. These activities, on the one hand, improve innovation performance; on the other hand, external networks may have substantial influences on these capabilities.

Hence, this chapter puts forward the following hypotheses:

H1a. KAC is positively related to innovation performance.
H1b. KGC is positively related to innovation performance.
H1c. KCC is positively related to innovation performance.

As explained earlier, there is an internal structure and indirect relationship between knowledge acquisition and generation and knowledge combination. However, the author proposes that knowledge combination capability contributes much more to innovation activities and performance. According to this indirect relation, the proposed model supposes that:

H2a. The relationship between KAC and innovation is mediated by KGC such that KAC is positively related to KGC.
H2b. The relationship between KAC and innovation is mediated by KCC such that KAC is positively related to KCC.
H2c. The relationship between KGC and innovation is mediated by KCC such that KGC is positively related to KCC.

The Antecedents of Dynamic Capabilities: Network Embeddedness

To investigate the effect of characteristics and structure of networks on the dynamic capabilities level of a firm, a term called "network embeddedness" is defined and taken as the antecedent of the dynamic capability. This chapter focuses on two dimensions called structural embeddedness and relational embeddedness. Structural embeddedness analyzes the structure of the integrated network system and focuses on the benefits it draws from the relative position in the network. Relational embeddedness emphasizes the characteristics of direct ties promoting deep and extensive knowledge exchange.

As the innovation process becomes more and more open and interactive, innovators must have the ability to grasp the variety of knowledge existing in their partner networks. When the network partners are more diverse, the focal firm will have more potential to get the required knowledge and to process innovative use of that knowledge. Based on the literature, this chapter argues that the nodal heterogeneity in the form of partner diversity will facilitate KDC:

H3a. Network diversity is positively associated with the KAC of the firm.
H4a. Network diversity is positively associated with the KCC of the firm.

Granovetter [11] emphasized the role of weak ties and pointed out that firms are more likely to get novel knowledge through weak ties. The philosophy is that the knowledge base between acquaintances may overlap greatly, leading to a high redundancy. Conversely, weak ties may transmit knowledge from totally different fields and inspire new ideas. Burt [12] joined this argument and pointed out that the literature should shift their focus from the strength of a tie to the overall structure of the network. Despite that, nonredundancy improves the opportunity of acquiring new knowledge; however, it implies diversity and a large volume of knowledge leading to a great challenge to knowledge combination. Hence, this chapter expects that:

H3b. Network nonredundancy will positively influence the KAC of the firm.
H4b. Network nonredundancy will negatively influence the KCC of the firm.

If firms feel confident about their partners, they will have positive expectations about the actions of their partners. This trust results both in a sharing and exchange of information between firms and partners and more confidence in the information they get. Hence, two hypotheses are proposed:

H5a. Trust will positively influence the KAC of the firm.
H6a. Trust will positively influence the KCC of the firm.

Sharing the responsibility to maintain the cooperation and to tackle the problems they confront during their cooperation with the network partners is called joint problem-solving. During this process, partners not only share explicit knowledge but also understand their partners' tacit knowledge and further promote knowledge exchange and assimilation. Hence, joint problem-solving is a platform for firms to experiment with different kinds of knowledge integration. The effect of joint problem-solving on dynamic capabilities is predicted as follows:

H5b. Joint problem-solving will positively influence the KAC of the firm.
H6b. Joint problem-solving will positively influence the KCC of the firm.

The existence and intensity of commitments are main elements in the sustaining and succeeding of alliances or partners' commitment from both firms discloses their knowledge pools leading to an increase in the knowledge-sharing and transferring levels. Hence, commitment will positively contribute to KDC:

H5c. The commitment will positively influence the KAC of the firm.
H6c. The commitment will positively influence the KCC of the firm.

METHODS AND MATERIALS

The empirical work (testing propositions) was based on a survey conducted during July 2015 to January 2016 in the Mazandaran industrial parks in the north of Iran.

The first questionnaire was developed through extensive literature research and fieldwork. However, this version was adapted according to the opinions of three kinds of experts. The content validity and the whole structure of the questionnaire were modified by consulting with corresponding researchers of this field. The first draft of the questionnaire was developed based on the literature, and then the survey was complemented by the persons most familiar with knowledge-related activities and alliance activities in each firm (chief technical officer, chief marketing officer and vice president).

1. Face-to-face investigation. Regarding the importance of this work, the author conducted a lot of field investigations and collected the survey.
2. E-mail. A manufacturing manager database consisting of managers' information from the Mazandaran industrial parks was formed in the central research center, the author sent the electronic questionnaire to the managers in this database through e-mail.

From 512 questionnaires sent to firms, 218 completed ones were used as the base for the hypotheses.

Dependent Variable: Innovation Performance

The following four proxies were used to reflect the innovation performance of various firms:

1. Number of new products;
2. Share of turnover with new products;
3. The speed of new product development and commercialization;
4. The ratio of successful product innovation.

Knowledge-Based Dynamic Capabilities (KDC)

Table 4.1 shows parallel instruments developed to measure the three components of KDC. Knowledge acquisition is a five-item instrument capturing the degree to which the focal firm could acquire technological, marketing, managerial, manufacturing and other relevant knowledge from its partners. Knowledge generation capability has the same structure as knowledge-acquisition capability and the instrument includes five items asking for the degree to which the firm could generate new technological, marketing, managerial, manufacturing and other relevant knowledge endogenously. Using the work of Grant [13], knowledge combination capability is measured with six items as shown in Table 4.1.

Independent Variables: Network Embeddedness

Following the work of [14], network embeddedness is defined as a two-dimensional construct: structural embeddedness measured by terms of nonredundancy and diversity, and relational embeddedness made up of trust, joint problem-solving and commitment.

Nonredundancy: This item is operationalized as an ego-centered network measure following the work of [14]. Using the respondents' information, based on the matrix of the five most important external relational partners and the interaction between each pair of partners, the redundancy score can be calculated using the following formula:

Nonredundancy = (number of potential ties – number of actual ties)/number of partners.

According to the formula above, the nonredundancy ratio varies between zero and two when five partners exist. Nonredundancy decreases if interactions between the focal firm's partners exist.

Diversity of alliances partners: This item was calculated and constructed based on the following relation:

$$\text{Diversity} = \frac{1-\left[\left(\sum \frac{i(\text{the number of alliance with } i^{\text{th}} \text{ type partner})}{\text{total number of alliances}}\right)^2\right]}{\text{total number of alliances}}$$

The diversity range varies between zero and one and increases as the diversity of alliances partners increases.

Trust: According to [14] following measures were used to evaluate this item.

1. The main partner fairly negotiates with alliance.
2. The main partner does not mislead alliance.
3. The main partner keeps its word.
4. The main partner is reliable.

Joint problem-solving: Using the work of [14], joint problem-solving was measured using the following items:

1. The main partner works with us to overcome difficulties.
2. We are jointly responsible with the main partners for getting things done,
3. We work with the main partner to help solving our problems.

Commitment: The following items were selected for this measure based on the work of [15]:

1. The firm is committed to this relationship.

TABLE 4.1
Measurement Instruments and Validity of Knowledge-Based Dynamic Capabilities

Construct	Measurement Items	Internal Consistency Reliability ($\propto$)	Standard Coefficients	CR
Knowledge acquisition capability	Our firm could acquire technological knowledge.	0.911	0.867	
	Our firm could acquire marketing knowledge.		0.830	13.362
	Our firm could acquire managerial knowledge.		0.817	13.014
	Our firm could acquire manufacturing and process knowledge.		0.790	12.289
	Our firm could acquire other knowledge and expertise.		0.773	
Knowledge. generation capability	Our firm could create technological knowledge.	0.938	0.817	11.218
	Our firm could create marketing knowledge.		0.906	12.800
	Our firm could create managerial knowledge.		0.875	12.233
	Our firm could create knowledge.		0.828	
	Our firm could create technological knowledge.		0.808	12.310
Knowledge combination capability	Our firm could combine internal and external knowledge.	0.938	0.891	14.425
	Our firm could integrate knowledge from different segments, teams, and individuals.		0.864	13.683
	Our firm could combine knowledge in different technological or market fields.		0.798	12.493
	Our firm could combine new knowledge with original knowledge pool.		0.879	12.309
	Our firm could adapt the internal structure and process to combine knowledge effectively.		0.851	13.359
	Our firm could coordinate internal and external networks to combine knowledge effectively.		0.841	13.115

Notes: $= \chi^2 = 120.026$; $\frac{\chi^2}{df} = 1.188$; $df = 101$; CFI = 0.991; TLI = 0.990; RMSEA = 0.035.

2. The firm makes maximum effort to maintain this relationship.
3. This relationship is something my firm intends to maintain indefinitely.

Control variables: According to [16], firm size was included as the control variable in this study. Firm size was measured as the average annual revenue of the two most recent years.

EVALUATIONS

Construct Validity and Reliability

Evaluating the reliability and validity of the constructs; focusing on the measures of the newly developed construct of KDC, the reliability of the KDC was examined with Cronbach's alpha. All scales have reliabilities greater than the recommended 0.90 level (shown in Table 4.1), suggesting high reliability. Regarding the criteria taken into account to carefully design the questionnaire, the content validity of this construct can be justified. Confirmative factor analysis was used to evaluate the convergent validity and discriminant validity that were above 0.77 and statistically significant, as shown in Table 4.1.

Hypothesis Testing

The hypotheses were tested using the method of structural equation modeling with the AMOS 7.0 software. This method can accommodate the measurement error of survey data. The result for the structural equation model is shown in Figure 4.1. The path coefficients and their significance in each hypothesis were reported. Several different indices such as χ^2 and χ^2/df were also provided to determine the overall fit of the estimated model. CFI and TLI indices are 0.927 and 0.920 (all above 0.9), and the value of RMSEA is 0.076 (below 0.08), these indices show that the estimated model has a reasonable fit with the data.

H1. The model fairly predicted and supported items included in H1a, H1b and H1c, although the coefficients and significance are lower than that of knowledge combination capability.

H2. The mediation effects within the three components of KDC, i.e., knowledge acquisition capability, knowledge generation capability and knowledge combination capability are largely supported by the data and the relationship between each of the components is significant and positive, see Figure 4.1.

H3. The hypothesized relationship between structural embeddedness and knowledge acquisition capability is not supported. Neither diversity nor nonredundancy showed any significant influence on knowledge acquisition capability, see Figure 4.1.

H4. According to Figure 4.1, the predicted relationship:

a. Between structural embeddedness and knowledge combination capability is partly supported.

b. Between diversity and knowledge combination capability is positive and statistically significant.
c. Between nonredundancy and knowledge combination capability is positive too, although not very statistically significant.

H5. The model states that relational embeddedness positively associated with knowledge acquisition capability and our result supported this hypothesis. As predicted, the relationship between trust and knowledge acquisition capability and between joint problem-solving and knowledge acquisition capability is positive and significant. The commitment contributes to knowledge acquisition capability positively, see Figure 4.1.

H6. The positive relationship between relational embeddedness and knowledge combination is largely supported. The result indicates that positive relationships exist between joint problem-solving and knowledge combination capability and between commitment and knowledge combination capability. However, the relationship between trust and knowledge combination capability is not supported.

Robustness of the Results

Some additional analyses such as testing the existence of the mediation effect were taken out. First, direct paths between network embeddedness and innovation performance were added. The coefficients for the additional paths are insignificant. Second, to test the mediation effects within the dynamic capabilities, three mediation paths were removed. Except for the link to knowledge generation capability, the direct path from knowledge acquisition capability is significant, at the 0.1 level. However, the overall fit indices get worse. These models offer consistent evidence to the theoretical model.

In this section, insignificant paths with the lowest CR value were trimmed off step by step resulting in a model in which all the coefficients were significant ($p < 0.05$). The comparison between the resulting model and the hypothesized one indicated that most significant paths were still significant except for some marginally significant paths and at the same time the explanative power of significant paths was enlarged. Hence, the best model gave us a more concise structure between the independent and dependent variables.

CONCLUDING REMARKS

This study extends and deepens our understanding of dynamic capabilities and its link with innovation performance in the networked environments by proposing a concise construct of dynamic capabilities and validating it through empirical studies. The resulting developed convergent construct can now be empirically tested. A clarification was made on conceptualizing dynamic capabilities aiming at addressing dynamic environments, their attributes and constitutions. This conception made it possible to identify and measure dynamic capabilities practically. This chapter offered rich evidence on the contribution of dynamic capabilities on innovation performance, systematically using an empirical study. Further, intrinsic structure within the three dimensions was discovered. Our results reported that knowledge combination capability promotes innovation performance directly while knowledge acquisition capability and knowledge generation capability contribute to innovation performance indirectly. These findings echoed the research of Kogut and Zander [17] that firms' innovations are products of their combinative capabilities. This study revealed that network embeddedness was an important antecedent of dynamic capabilities, greatly influencing knowledge acquisition capability and knowledge combination capability. The results of our empirical study indicate that relational embeddedness exhibits a greater influence on the KDC. Knowledge acquisition capability is mainly influenced by trust, joint problem-solving, and commitment, while knowledge combination is mainly driven by joint problem-solving and commitments. An interesting result was that nonredundancy has no significant links with dynamic capabilities, which does not align with the literature.

Despite all the findings made by this research, further questions emerged in this study. First, the construct of KDC calls for more examination and verification. The conceptualization and operationalization of dynamic capabilities need more empirical studies based on large samples, different contexts and much work remains to be done to yield a mature construct. Second, this chapter does not examine the differences between different industries' high-tech sectors and stable industries such as the steel industry. Adding control variables or restricting their investigations to specific kinds of industry in further studies may overcome this problem. Third, there is a possibility that firms will reconstruct their organizational environments deliberately to sustain competitive advantage (the evolutionary nature of dynamic capabilities). In other words, managers may adapt their alliance networks dynamically. Hence, feedbacks between network environments, dynamic capabilities and innovation performance link the cycle of these constructs.

REFERENCES

1. Teece DJ, Pisano G, Shuen A. Dynamic capabilities and strategic management. *Strat Manag J.* 1997;18(7):509–33.
2. Helfat CE, Finkelstein S, Mitchell W, Peteraf M, Singh H, Teece D, Winter SG. *Dynamic capabilities: understanding strategic change in organizations.* Chichester: Wiley-Blackwell; 2007.
3. Zollo M, Winter SG. Deliberate learning and the evolution of dynamic capabilities. *Org Sci.* 2002;13(3):339–51.
4. Teece DJ. Explicating dynamic capabilities: the nature and microfoundations of (sustainable) enterprise performance. *Strat Manag J.* 2007;28(13):1319–50.
5. Teece DJ. *Dynamic capabilities and strategic management: organizing for innovation and growth.* Oxford: Oxford University Press; 2009.
6. Lane PJ, Koka BR, Pathak S. The reification of absorptive capacity: a critical review and rejuvenation of the construct. *Acad Manag Rev.* 2006;31(4):833–63.
7. Nonaka I. A dynamic theory of organizational knowledge creation. *Knowl Manag Crit Persp Bus Manag.* 2005;5(1):14–37.
8. Wadhwa A, Kotha S. Knowledge creation through external venturing: evidence from the telecommunications equipment manufacturing industry. *Acad Manag J.* 2006;49(4):819–35.

9. Hoffmann WH. Strategies for managing a portfolio of alliances. *Strat Manag J.* 2007;28(8):8–27.
10. Lavie D. Alliance portfolios and firm performance: a study of value creation and appropriation in the US software industry. *Strat Manag J.* 2007;28(12):1187–212.
11. Granovetter M. Economic action and social structure: the problem of embeddedness. *Am J Sociol.* 1985;91(3):481–510.
12. Burt RS. *Structural holes: the social structure of competition.* Cambridge, MA: Harvard University Press; 1992.
13. Grant RM. Prospering in dynamically-competitive environments: organizational capability as knowledge integration. *Org Sci.* 1996;7(4):375–87.
14. McEvily B, Marcus A. Embedded ties and the acquisition of competitive advantage. *Strat Manag J.* 2005;26:1033–55.
15. Morgan RM, Hunt SD. The commitment-trust theory of relationship marketing. *J Mark.* 1994;58(3):20–38.
16. Shefer D, Frenkel A. R&D, firm size and innovation: an empirical analysis. *Technovation.* 2005;25(1):25–32.
17. Kogut B, Zander U. Knowledge of the firm: combinative capabilities, and the replication of technology. *Org Sci.* 1992;3(3):383–97.

5 Knowledge and Organizational Business Loss

RESEARCH SCOPE

This work suggests a framework for the effective performance of knowledge retention, also termed vertical knowledge transfer, the middle of three states an organization can experience. The first includes organizations that intensively and regularly manage their knowledge. These avoid the need for extensive knowledge retention when people leave. The second is part of the "the day after" organizations, which are not included in our research. The third state includes organizations that neither thwart the need in advance nor take any action for knowledge retention when people leave. These organizations will find themselves reacting to knowledge loss after knowledge loss. Knowledge retention is an action for organizations between these states that start planning their knowledge retention process 3–12 months before retention takes place. This work focuses on the intermediate state of first and third states of knowledge retention.

LITERATURE REVIEW

Before suggesting a framework for knowledge retention, the existing literature was studied. Knowledge retention projects should include three stages:

The first stage deals with the decision-making – whether and at what level vertical knowledge transfer is required. The second stage includes the planning – defining the knowledge to be retained and how. The last stage deals with the practical implementation of the plan (see Figure 5.1).

The decision is based on the awareness of the risk of knowledge loss and assessments as to the problem's magnitude. There is the challenge of convincing managers that a problem of knowledge loss exists. Some are aware of the phenomenon but are not aware that it is a problem. Once organizations have understood, they decide to act, while others try to further understand the magnitude of the problem by conducting a double level assessment. This involves assessing probable retirees by learning age distribution through the human resources (HR) computing system [1,2].

Planning is an important stage including two main phases:

1. Determining the knowledge to be vertically transferred, resulting in a list of names of those retiring or leaving, departments, jobs and roles at risk and hence, the suggested knowledge to be transferred; and

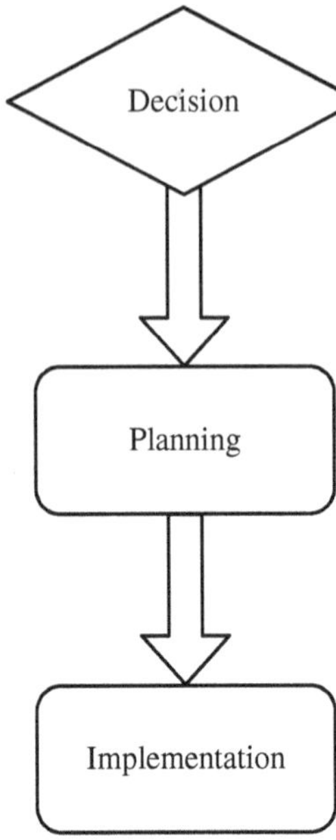

FIGURE 5.1 The general knowledge retention stages.

2. Determining how to transfer the knowledge and retain it within the organization. Due to the infinite nature of the knowledge held by those leaving and to limited resources of time and money, the knowledge to be retained has to be prioritized.

Vertical knowledge transfer, as described in this research, transforms the knowledge into part of the organization's assets. The viable solutions are such as the use of a document database with eased access, including standard procedures, monthly reports, regulations, drawings, maps, manuals and annual analysis.

Researchers suggested 12 strategies: job shadowing; communities of practice, process documentation, critical incident interviews or questionnaires, expert systems, electronic performance support systems (EPSS), job aids, storyboards, mentoring programs, storytelling, information exchanges and best practice studies or meetings [3,4].

DeLong [3] also suggests eight strategies in knowledge management (KM) of which three of the strategies aimed at improving the transfer of explicit knowledge are: documentation, interviews and training – and four for transferring implicit and tacit knowledge – storytelling, mentoring/coaching, after-action reviews and communities of practice. They refer to two other solutions that bypass the problem: eliminating the need for the knowledge by changing processes or equipment, etc., or establishing alternative resources like outside contractors or retirees as consultants. As tacit knowledge is difficult to transfer, one way could be recreating tacit knowledge.

Several potential candidates are responsible for the vertical knowledge transfer process. Beazley sums up the list: "Such authority might be given to the chief information officer, the chief knowledge officer, the human resources director or an individual appointed to coordinate the start-up" [5].

Key Success Factors Described in Literature

Slagter [6] offered critical success factors of KM tailored for work with senior employees. Her findings related to knowledge retention: a coaching leadership style, granting the senior employees responsibility and an active role in job transfer, attention to motivation, trust, rewards and recognition and mixed teams and work groups containing both senior and junior employees. According to DeLong the relationship between young and the old value knowledge and poor communication between different generations is the main barrier to knowledge retention.

Literature Critique

For both the academic studies and the case studies most emphasis is placed on the first stage of knowledge retention. The author suggests that an assessment project is unnecessary and suggests skipping this stage and starting instead with a modest pilot. The secondary criticism deals with the process of knowledge transfer itself. As most works place most emphasis on the assessment, few describe in detail the following stages, which are the core of the process. A new framework is integrating ideas from various works in the literature and their knowledge management experience which is tested and indicates positive results [7].

Lessons Learned

If the organization and management required perception, there is no need for assessment, and it is recommended to skip this stage saving money, time and human-based efforts. Templates were found to be very helpful in guiding the capturing process. These served well also in easing reuse of knowledge by the incomers. While organizing documents that already existed, it was found helpful to add summaries to each important document; adding summaries would be helpful because new incomers tended not to open documents. The documented knowledge was integrated into the IT organizational environment. Even if there is not such an infrastructure, it can be done by the WIKI system on the internet [8].

PROPOSED FRAMEWORK

Three levels exist for managing the knowledge continuity of an organization as shown in Figure 5.2: avoidance, engagement and reaction.

Organizations that regularly manage their knowledge avoid the risks faced by most organizations when their employees leave. These will not usually put the organizational knowledge or the organization's business competitive advantage at risk. A reaction is required when an organization faces the loss of knowledge after

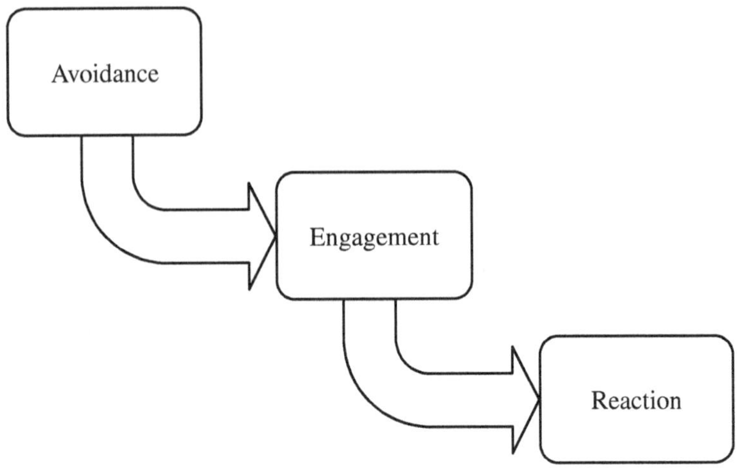

FIGURE 5.2 Managing knowledge continuity.

employees have left with that knowledge, leaving the organization with a business loss. The organization must then discard the need for the lost knowledge or buy knowledge services to fill in what is missing. This is achieved through consultants, outsourcers or, in some cases, re-employment of the retirees. The suggested framework focuses on the middle level: engagement. It offers organizations facing the risk of significant knowledge loss, a process for engaging the issue and vertically transferring the knowledge. The basic assumption does not suggest that all knowledge can be retained; in all cases, some knowledge will be lost; the loss, however, should be less significant to the future business.

RESEARCH METHOD AND RESULTS

The research conducted was based on a multi-case study methodology. Yin [9] counted five different research methods: experiment, survey, archival analysis, history and case study. Choosing this specific method should be based on three conditions: form of the research question, control of behaviors and events and focus on contemporary events. The only relevant method is the case study method. The multi-case method enables richer research, where the generalization of the results is more reliable than when only analyzing one case. Yin [9] defined five question components to be included in a case study research design:

1. Study questions;
2. Its propositions, if any;
3. Its unit(s) of analysis;
4. The logic linking the data to the propositions; and
5. The criteria for interpreting the findings.

The research of knowledge retention was designed considering these issues.

The knowledge retention research question (component 1 above):

How can organizations minimize the loss of important knowledge while experiencing high levels of retirees?

The knowledge retention research proposition (component 2 above):

The research proposes a framework that has been shown to improve knowledge retention processes in organizations. The suggested framework emphasizes a structured process with structured results and predefined anchor points by which the knowledge will be reused within the organization [10].

The knowledge retention research unit of analysis (component 3 above) is the organization. Each organization in which a retaining project was performed, where people whose knowledge was to be retained, were chosen, and where the project was documented, was defined as a unit of analysis. While designing the research, the question arose as to whether each retiree should be a separate unit of analysis, or if was it more suitable to define the whole organization as one unit. Sometimes each retiree is a full project regarding carrying on separately, the stages of scope definition, knowledge transfer, and knowledge integration. On the other hand, in some departments, projects were initiated with two retirees, and the preliminary stage was united [11]. The preliminary stage and the full surroundings are on an organizational level, the core of the methodology is personal.

The data were collected speaking both with knowledge managers and business managers in the organizations where the projects were conducted. It must be noted that the organizations paid for the retention projects and were not aware while running the projects themselves. In the knowledge retention research described, data were linked to the propositions and interpreted using the explanation building technique including pattern matching, explanation building, time-series analysis, logic models and cross-case synthesis as serving to be more suitable for the case. The research did not include the full framework in advance. Possible techniques for retaining knowledge were listed beforehand, based on a literature review and brainstorming.

The suggested framework includes four stages to be carried out in knowledge retention projects. These four stages are necessary, and none can be skipped, offering a shorter methodology. The preliminary stage focuses on the initiation of the project. The core stage of transfer, including planning and implementation, is the heart of the project and, of course, cannot be skipped. The final step is knowledge integration, facilitating the knowledge utilization and reuse for the future. This step has been shown to be essential. When it was skipped, people tended not to use the past knowledge as it was not sufficiently accessible and they were not reminded or even obligated to do so (see Figure 5.3).

In all stages, special care is given to the structuring of the process and the results. Structuring the process promises project progress and efficiency; structuring the results improves knowledge understanding, and therefore improves the project's effectiveness.

The preliminary stage focuses on initiating a project and conducting it on an organizational level, although special people are considered to retire. The steps are usually carried on separately for each retiree. The first stage defines the project's scope

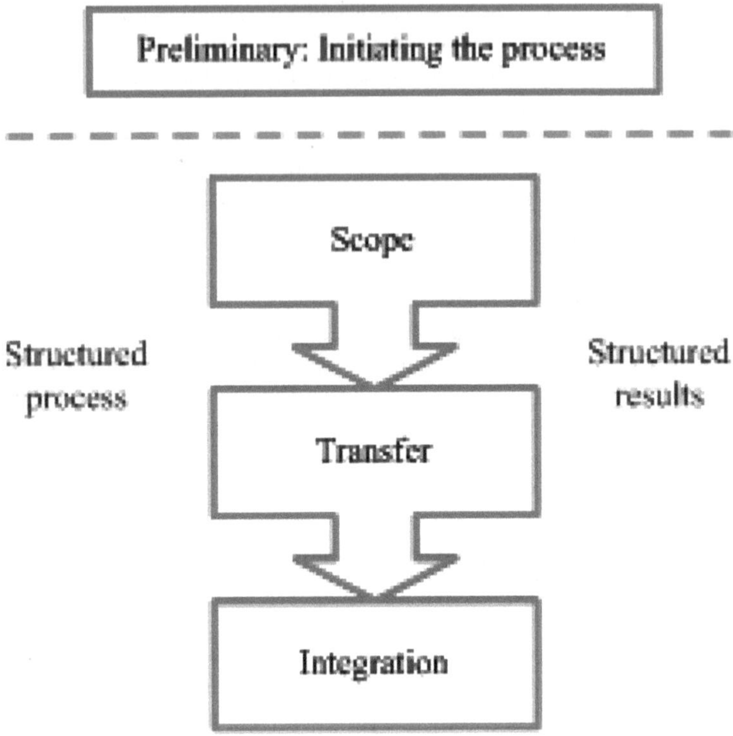

FIGURE 5.3 The proposed knowledge retention stages.

– what knowledge will be retained and what will be skipped. While the concepts of knowledge retention were introduced, some managers did not accept the practicability of knowledge retention.

The first preliminary step in most organizations analyzed included: introducing the retention and knowledge retention issue, bringing it to the awareness of the high-level managers and attracting departments to request this service. In all cases, the first department was very easily located, and pilot projects were defined; accordingly, the immediate managers pointed out the potential retirees holding valuable knowledge.

The suggested framework was designed including a preliminary stage of introducing the knowledge retention issue, raising the management's awareness of the possibilities and success of the methodology implemented in other organizations and yielding a pilot of a few first retirees with whom, or areas in which, knowledge retention is to take place. Once such pilots have been conducted, it became much easier to enlarge the project's scope, adding potential departments, areas and retirees.

This stage is preliminary to the project itself and should not consume significant resources. To prevent an unsuccessful choice of retirees, the managers should be guided regarding what to consider when focusing; intuitive decisions are validated, and extreme cases where knowledge retention is not applicable are eliminated.

Skipping this stage is risky; it may yield a project that transfers much more knowledge and therefore may result in being too long, and too expensive. More importantly, as there is hardly ever the time to pass all, people may retire, leaving significant knowledge outside the organization, causing business loss. People tend to transfer what is easy.

A simple technique is based on building a knowledge tree, which includes all knowledge subjects, and explicitly marking the branches to be retained and those to be discarded. For each knowledge subject marked for retention, a description should be added regarding the reason this knowledge is important.

The scoping step ends when the manager signs this list (or tree). The manager should understand the system as a whole and consider not only the present aspects but also evaluate the future ones while evaluating importance, as prioritization is unique, compared to similar suggested processes.

Transferring the knowledge from the retiree into the organization stage refers to documented knowledge and that knowledge, explicit and tacit, is only possessed by the retiring employees, whether in their minds or the computers. Documented knowledge will be shared. The project will include gathering and organizing existing documentation of the retiree/subject and storing it in a shared location.

While planning and implementing the transfer stage, it is recommended to further refine prioritization of the more valuable knowledge within the selected subjects for retention, thus, filtering the knowledge to be retained with future needs in mind.

Hofer-Alfas suggests four types of knowledge structures used for capturing and transferring the knowledge:

1. The knowledge area list, including business-critical areas of proficiency;
2. The knowledge relationship map, including business-critical relationships and networks;
3. The knowledge portfolio, including business-critical codified knowledge assets, e.g., records and instruments;
4. The lessons learned table.

There may be various formats of knowledge documentation, starting with documents, through videos, exposing meetings, embedding knowledge items in knowledge bases, etc. Deciding on a documentation format depends on three factors: the epistemological type of knowledge, the complication level of the knowledge and the existing formats of knowledge within the organization and specific area. The complication level of knowledge is another factor since complicated knowledge is better transferred when people can listen and not only read, enabling them to ask and request clarification or reflect on what they have understood. Documentation storage also must be planned. It may be based on a portal system, a document management system or any other knowledge management infrastructure used in the organization.

The retaining process is often the first tangible process that drives home the reality of retirement to the person leaving. Many people fear this non-reversible change, and their feelings are reflected in their reaction towards those requesting their knowledge. Knowledge retention project managers should be aware that this is not a mere technical process and should send sensitive people to the task, people who can handle these fears or even assist in relieving them. The transfer stage has to be controlled and is reviewed by time and budget.

How the transferred knowledge is integrated and embedded into organizations' processes is significant. Knowledge retention is of little benefit if the knowledge is not used by the organization after retirement. In some cases, knowledge was integrated into routines and business processes in which it was likely to be used. This is the core of knowledge integration: analyzing scenarios and anchor points in which the knowledge can be linked and creating these connections so that people will easily be exposed to the knowledge where and when it is needed.

CONCLUDING REMARKS

The purpose of this chapter was to minimize knowledge loss in organizations.

Knowledge retention is an important concept recently developed by researchers. Studies have been conducted and described in the literature, mainly suggesting three stages to be followed when aiming to retain knowledge: decision, planning and implementation. The present research, based on seven case studies (each with two to 14 retirees) between the years 2014 and 2017, showed that organizations found it easy to understand the need for knowledge retention once they were aware of the concept, even though no preliminary assessment project was conducted. Knowledge retention has been an important issue occupying researchers in recent decades. They mainly suggest three stages to be followed when aiming to retain knowledge: decision, planning and implementation. Organizations found it easy to understand the need for knowledge retention once they were aware of the concept. Projects, conducted in seven organizations, succeeded with no assessment, and six of them yielded a larger knowledge retention project. In one other case, where an assessment project did take place, the pilot stage was not initiated since all of the budget and time were spent on the assessment project. Transferring the knowledge by organizing and sharing existing documentation and by documenting undocumented knowledge promises the reuse of the retained knowledge. In the whole of the process, the emphasis should be on structuring the process and results.

REFERENCES AND FURTHER READING

1. Dalkir K. *Knowledge management in theory and in practice.* Burlington, MA: Elsevier Butterworth-Heinemann; 2005.
2. Levy M. Leveraging knowledge understanding in documents. *Elec J Knowl Manag Prac.* 2009;7(3):57–69.
3. DeLong DW. *Lost knowledge.* New York, NY: Oxford University Press; 2004.
4. Leonard D, Swap WC. *Deep smarts.* Boston, MA: Harvard Business School Publishing Corporation; 2005.
5. Hofer-Alfeis J. Knowledge management solutions for the leaving expert issue. *J Knowl Manag.* 2008;12(4):44–54.
6. Slagter F. Knowledge management among the older workforce. *J Knowl Manag.* 2007;11(4):82–96.
7. Hofer-Alfeis J. Leaving expert debriefings to fight the aging workforce's retirement wave. *I J Hum Res Dev Manag.* 2009;9(2/3):300–04.

8. Kalkan VD. Knowledge continuity management process in organizations. *J Bus Eco Res.* 2006;4(3):41–46.
9. Yin RK. *Case study research: design and methods.* 4th ed. Newbury Park, CA: Sage Publications; 2009.
10. Davenport TH. *Thinking for a living: how to get better performance and results from knowledge workers.* Boston, MA: Harvard Business School Press; 2005.
11. Fazlollahtabar H, Muhammadzadeh A. A knowledge-based user interface to optimize curriculum utility in an e-learning system. *Int J Enterp Inf Syst.* 2012;8(3):35–54.

6 Knowledge Sharing Using Semantic Web

INTRODUCTION

An efficient information/knowledge sharing approach is more important than ever in tightly coordinating these independent entities (partners or companies) in a supply chain and preventing the bullwhip effect. However, only a few studies have focused on the interoperability problem of knowledge sharing. When knowledge workers interact with unfamiliar knowledge sources that have been independently created and maintained, they must make a non-trivial cognitive effort to understand the information contained in the sources. Moreover, information overload makes achieving semantic interoperability between a knowledge source and a knowledge receiver (such as a designer, decision maker and an application or peer agent) more critical than ever. Sheth [1] classified three types of interoperability problems, including system, syntactic, structural and semantic levels of heterogeneity.

Problems of interoperability among interacting computer systems have been well documented. Current technologies, for example, the internet and the world wide web (WWW) support a dynamic and unprecedented global information infrastructure for supply chains. However, one of the major problems is the huge amount of information available and our limited capacity to process it. However, the main problems that arise concern system, syntactic or structural issues, but an important aspect of information/knowledge – meaning (semantic) – has not been properly addressed. Although such schema-level specifications can be used successfully to specify an agreed set of labels with which information can be exchanged, the current web technology cannot be assumed to solve all problems of semantic heterogeneity.

A summary of interoperability problems in knowledge sharing:

1. Too many such schema-level specifications exist, and they may not all use the same terminology.
2. Bussler [2] showed that business to business (B2B) integration through programming does not scale, because of the high complexity of the interactions of B2B protocols such as electronic data interchange (EDI) or RosettaNet.
3. The current technologies, for example, EDI or RosettaNet, do not explicitly link the semantic requirements to formal process models. The missing links make infeasible the integration of supply chain management (SCM) implementations.
4. The problem of semantic heterogeneity still applies while all data are exchanged using Extensible Markup Language (XML), structured according to standard schema-level specifications.

Therefore, a potential solution that involves semantic technologies is required in a supply chain; such technologies have the potential to revolutionize the IT world. Semantic technology, for example, the semantic web has attracted much interest and has been applied in many areas. In a world of heterogeneous information, the semantic web enables flexible and seamless integration of applications and data sources. The semantic web provides intelligent access, an understandable context, and inferred knowledge. Furthermore, it provides a well-defined structure, ontology, in which meta-knowledge can be applied. The semantic web has great potential to share knowledge in a scalable manner [3].

MAIN BODY

Solution approaches to knowledge sharing involve: (1) a semi-structured knowledge (SSK) model; (2) an agent-based annotation process; (3) an articulation mechanism, given two used ontologies (benchmark ontology and other heterogeneous ontology).

Semi-structured knowledge (SSK) model: The knowledge is hard to manage and store, because of its structural uncertainty. The knowledge document is formulated using the SSK model, based on the six dimensions of the Zachman Framework: who, what, when, where, why and how. The SSK model has been formulated by the processes of knowledge production:

- Classification of events;
- By definition;
- The general description of the event;
- The feedback process and conversion of the notes or annotations;
- Links between documents and knowledge in the semi-structured;
- Semi-structured knowledge for the annotation process;
- Semi-structured knowledge.

Semi-structured knowledge is generated as follows:

$$k = k_o \cup k_A$$

$$k_o = I_G \cup k_P \cup k_T$$

$$k_A = D_B \cup D_A \cup D_R$$

k_O: organizational semi-structured knowledge;
k_A: semi-structured knowledge after annotation;
I_G: general information about the problematic event;
k_P: knowledge derived from knowledge workers;
k_T: the knowledge captured from the experts' feedback;
D_B: the basic description of annotated knowledge;
D_A: descriptive information about the annotation;
D_R: description of the relation among the documents, ontology, and generated knowledge.

Figure 6.1 illustrates the contents of final SSK in the XML format.

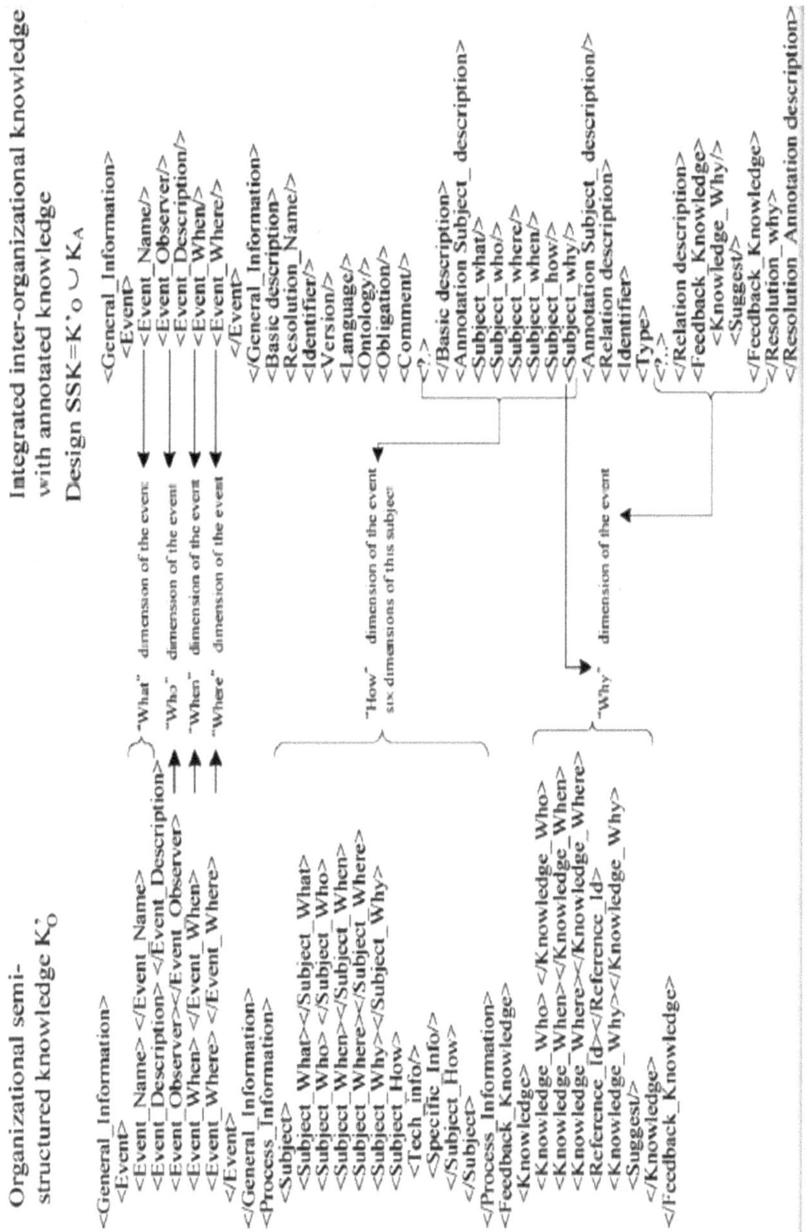

FIGURE 6.1 Example of SSK documentation.

The left side of Figure 6.1 presents general information (I_G), and feedback from knowledge workers (k_T), where the association rule algorithm is applied as a solution approach, and the right side of the figure presents the composition of SSK.

The annotation process relates to the unstructured formats of distributed knowledge documents.

D_A is based on the 5W1H (what, where, who, when, why and how) to represent annotated knowledge using the SSK model.

D_B includes general statements about the knowledge document and its source.

D_R presents two types of relationship – (i) the relationship between the annotated knowledge document and other heterogeneous/unstructured documents, and (ii) the relationship between the benchmark ontology and heterogeneous ontology.

According to the annotation process in Figure 6.2, the knowledge documents in heterogeneous formats are annotated and stored as follows:

Step 1: The annotation agent extracts the knowledge contents from unstructured or heterogeneous knowledge documents. The annotation agent is responsible for inserting XML-description tags into knowledge contents. The XML-based metadata for these knowledge contents is generated.

Step 2: The annotation agent is responsible for identifying metadata of knowledge contents as (D_B; D_A; D_R) and for transforming them into Resource Description Framework (RDF) and RDF Schema (RDFS) format documents. The annotated knowledge (k_A) represented using the SSK model is generated. Conception schemas are also created in this step. The perception schema is formed in an RDFS file and describes the structure of the concepts and properties for the annotated knowledge.

Step 3: (1) The annotation agent stores the annotated knowledge document together and the conception schema in the metadata repository. (2) The annotation agent registers the link to the original document in the register.

Articulation

Articulation is a mechanism for supporting the interloper ability of various sources and solves the problems of semantic interoperability. The articulation mechanism proposed in this section (i) efficiently resolves the heterogeneity of ontologies utilized by various business entities and (ii) appropriately intersects, unifies and differentiates heterogeneous ontologies with the benchmark ontology, according to purpose – for example, "intersection" for clarifying the interoperating part of the two ontologies and "difference" for validating articulation rules, respectively [4].

Knowledge-Sharing Platform

The proposed platform comprises (1) an agent-based annotation component, (2) an articulation component, (3) an interface component to present the SSK model, (4) a knowledge-based component to store the benchmark ontology and (5) a transfer component to transfer knowledge through the semantic web [5].

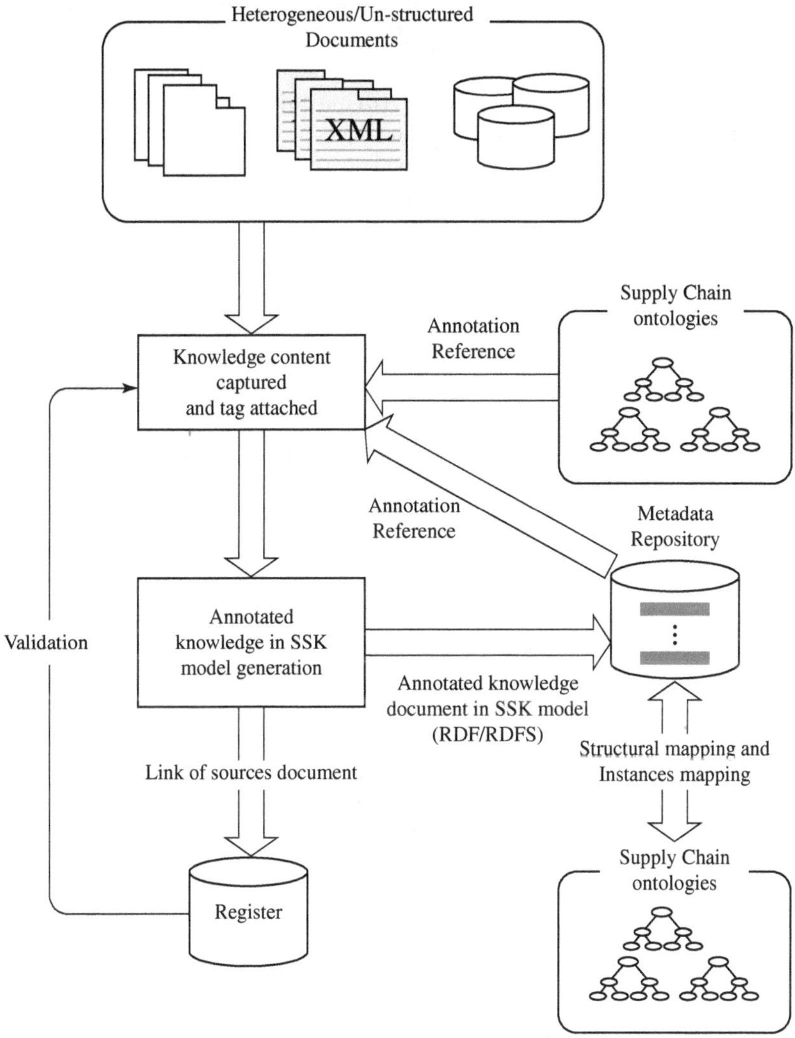

FIGURE 6.2 Annotation process.

An agent-based annotation component enables the annotation agent to receive unstructured and heterogeneous format documents from heterogeneous knowledge sources. Second, the knowledge contents of these documents are extracted, and the useful knowledge contents of these documents are annotated using description tags. Third, the annotated knowledge contents are organized into three annotation categories: (i) basic description of annotated knowledge D_B, which includes basic information about the problematic event, (ii) descriptive information about annotation D_A, which details an annotated subject and (iii) description of relation among the documents, ontology, and generated knowledge D_R; which describes the relationships of annotated knowledge (including articulation information) with the benchmark ontology and original documents. Fourth, these descriptions are transformed into RDF and RDFS formats. Accordingly, the annotated knowledge k_A is generated. Finally, (i) the annotated knowledge document representing and using the SSK model is stored in the metadata repository; (ii) the link to the location of the original heterogeneous document is registered in the register repositories [6].

The articulation component is to provide a scalable and simple framework to develop articulation. Interface component has three sub-components – the requesting and presenting (R&P) interface sub-component, the modeling and generating (M&G) sub-component and the knowledge querying and offering (KQO) sub-component. The R&P sub-component requests and presents semi-structured knowledge interacting with the M&G sub-component. The M&G sub-component is used to model and generate SSK. The ontologies are represented using a conceptual graphic for domain experts to articulate. KQO sub-component queries (i) knowledge documents in the SSK model stored in the knowledge base; (ii) annotated knowledge in the SSK model stored in the meta-data repository and (iii) knowledge elements using in the benchmark ontology, providing them to the M&G sub-component [7].

The knowledge-based component's main purpose is to store elements using in the benchmark ontology, articulation rules and knowledge documents in the SSK model, where (i) concepts and relationships between concepts are stored in RDFS format and (ii) instances of these concepts with corresponding values of properties are stored in RDF format. In this component, the "message exchange and search" agent presented is involved in (i) seeking the desired knowledge files in every supply chain entity's metadata repository, register and knowledge base and (ii) transferring these desired knowledge files to knowledge workers via the semantic web or simply through the general web [8].

Figure 6.3 summarizes the knowledge-sharing process. Knowledge workers and application agents seek knowledge for resolving problematic events. The knowledge-sharing platform supports the retrieval of the desired SSK. The steps are detailed as follows:

Step 1: Knowledge workers illustrate a problematic event.

E, I_G using the SSK model offered by the interface component. Peer agents can also acquire knowledge using this interface. The interface component sends the requirements, which are filled in by the peer agents, to request and search for desired knowledge elements, and aids knowledge workers to illustrate problematic events with these elements [9].

Step 2: If (i) sought elements of particular SSK are present in the benchmark ontology, (ii) sought knowledge documents in the SSK model are present in the knowledge base or (iii) a useful knowledge document link is present in the register and the knowledge has been annotated and stored in the metadata repository; then these elements and these documents are sent directly to users and agents. Otherwise, the conversation process in step 3 is initiated.

Step 3: Initiate the conversation to acquire the desired knowledge. Since the links between knowledge sources are all registered in the register repository, the required knowledge in the registered knowledge source can be

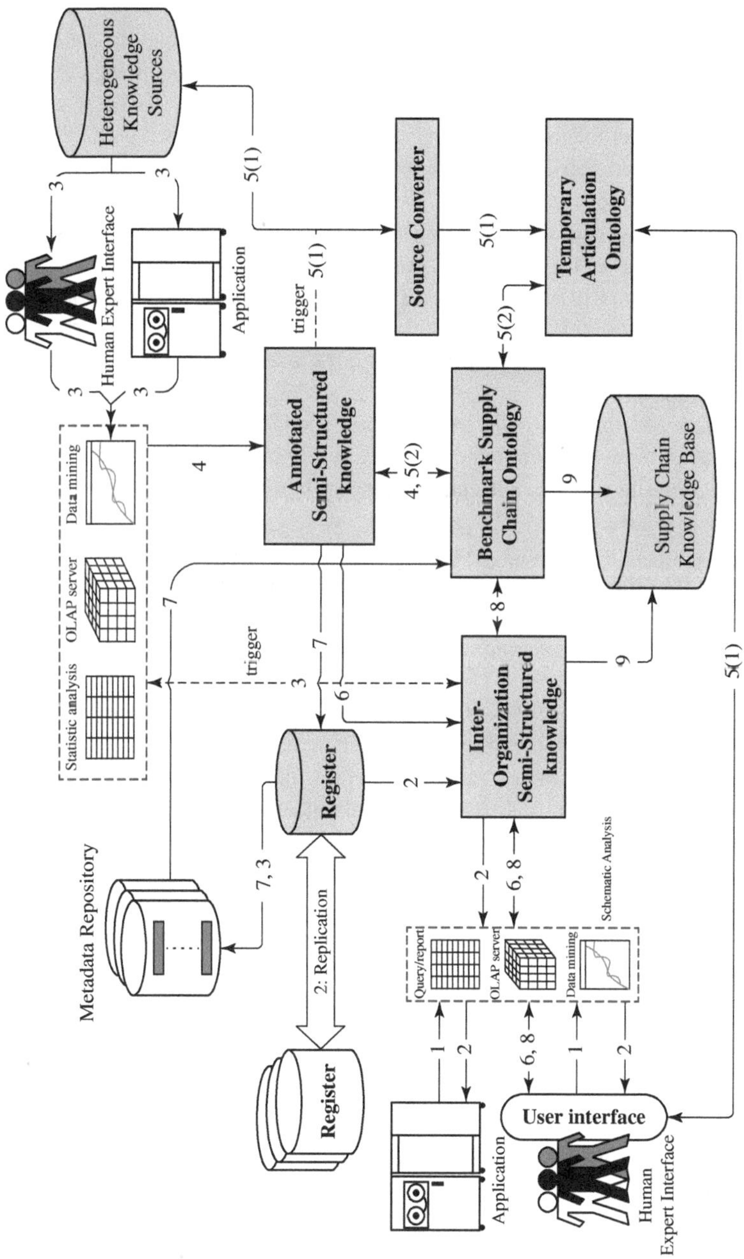

FIGURE 6.3 A summary of the sharing process.

extracted. During this process, the search agent, employed in the semantic web technique [10], constructs an inter-organizational communication bridge and delivers heterogeneous knowledge documents to the knowledge workers according to their requirements.

Step 4: In this annotation process, the annotation agent annotates unstructured/heterogeneous knowledge documents and generates k_A using the SSK model in the form of "annotated knowledge documents". If the original knowledge documents are obtained from hetero-partners, which have their ontology and compose knowledge in their specific format, then go to step 5, the articulation process.

Step 5(1): In this step, the heterogeneous ontology of hetero-partner is represented using a graphical conceptual model. Then, the articulation generator uses the heuristic methods to identify articulation rules. Temporary articulation ontology is constructed using ontology algebra. It is presented and browsed by experts to validate the articulation rules by the interface component.

Step 5(2): Using these articulation rules, knowledge documents from heterogeneous sources are annotated as annotated knowledge in the SSK model (k_A). After the articulation ends, the articulation rules are stored in the knowledge base.

Step 6: The document that includes annotated knowledge (k_A), is simply browsed via the R&P interface sub-component only, or it further adds other SSK.

Step 7: In this step, the annotated knowledge documents are stored in the meta-data repository, and the index and links between the locations of knowledge sources are registered. Each entity in the supply chain has a register. They all upgrade and replicate information on links synchronously; therefore, the annotation component always maintains consistent content. After that, annotated data (metadata) about annotated knowledge documents are extracted and mapped onto the benchmark ontology.

Step 8: The M&G sub-component presents the subject of the problematic event and its solution in the SSK model to experts and knowledge workers to resolve the problematic event. All knowledge workers are allowed to give feedback solution-related knowledge.

Step 9: Store the SSK documents and the updated benchmark ontology in the knowledge base.

CONCLUDING REMARKS

This chapter proposed the solutions for sharing knowledge in a supply chain, specifically solving the problem of knowledge interoperability; (i) the SSK model was presented to formulate knowledge not only to be explicit and sharable, but also meaningful; (ii) the annotation process was presented to annotate heterogeneous and unstructured knowledge documents as k_A and was stored in the meta-data repository; (iii) the articulation mechanism was presented to resolve semantic heterogeneity between two ontologies; (iv) the semi-structured knowledge-sharing platform was presented and based on the semantic web. The platform allowed the entities in the supply chain to represent, generate, search for and share knowledge effectively, and (v) the knowledge sharing process clarified these solution activities.

REFERENCES

1. Sheth AP. *Changing focus on interoperability in information systems: From system, syntax, structure to semantics*, M. F. Goodchild, M. J. Egenhofer, R. Fegeas, & C. A. Kottman (Eds.) (pp. 5–30), Boston: Kluwer, Academic Publishers, 1998.
2. Bussler C. Modeling and executing semantic B2B integration. In: *Proceedings of the 12th International Workshop on Research Issues in Data Engineering: Engineering e-Commerce/e-Business Systems (RIDE'02)* (pp. 69–74). San Jose, CA; 2002.
3. Huang CC, Kuo CM. Transformation and searching of semi-structured knowledge in organizations. *J Knowl Manag.* 2003;7(4):106–23.
4. Malone D. Knowledge management a model for organizational learning. *I J Account Inf Sys.* 2002;3(2):111–23.
5. Premkumar G. Inter-organizational systems and supply chain management – An information processing perspective. *Inf Sys Manag.* 2000;17(3):56–69.
6. Stefansson G. Business-to-business data sharing: A source for integration of supply chains. *Int J Prod Econ.* 2002;75(1–2):135–46.
7. Zhuge H. A knowledge grid model and platform for global knowledge sharing. *Exp Sys Appl.* 2002;22(4):313–20.
8. Chen MC. *Knowledge sharing in supply chain with semantic web.* Dissertation. Taiwan: Department of Information Management, National Chi-Nan University; 2003.
9. Huhns MN., Stephens LM. Automating supply chains. *IEEE Int Comp.* 2001;5(4):90–3.
10. Finin T, Joshi A. Special section on semantic web and data management: Agents, trust, and information access on the semantic web. *ACM SIGMOD Rec.* 2002;31(4):75–88.

7 Knowledge Sharing and Tax Payment

INTRODUCTION AND BACKGROUND

One way to finance government expenditures is to collect taxes. Regarding this financial source compared with other sources, positive sharing of tax knowledge amongst people or taxpayers leads to effective investment. Unlike developing countries, where taxes have little effect, in developed countries almost all government expenditure is financed by taxes. One of the main challenges in the tax system is how to collect the taxes lost due to tax evasion. The main reason is the uncertainty surrounding how the government uses the taxes paid by the people. A major factor in cases of failure to pay taxes is the discussion and sharing of views. If there is any perception of a positive tax effect, it motivates them to pay more, and if not, the paying of taxes is impaired. Therefore, to avoid disorderliness in paying taxes, which leads to a reduction in the development growth rate of investing taxes in industry and the services sector, procedures should be designed so that the positive effect of taxes becomes more widely discussed. In this chapter, five categories of how people share their knowledge about the benefits of taxation, have been proposed. Defining risk structure and using data from surveys, form the risk values of tax payment. The results indicate that sharing tax knowledge amongst people has positive effects on tax payments. Tax is defined as "a compulsory levy, imposed by government or other tax-raising body, on income, expenditure or capital assets, for which the taxpayer receives nothing specific in return" [1]. Taxes can be classified into two main types: direct and indirect taxes. Direct taxes mean the burden (incidence) of tax is borne entirely by the entity that pays it and cannot be passed on to another entity, for example, corporation tax and individual income tax. Indirect taxes are typically the charges that are levied on goods and services (consumption), for example, VAT (Value Added Tax), sales tax, excise tax and stamp duties [2].

For the development and growth of any society, the provision of basic infrastructure is a necessity. This perhaps explains why the government shows great concern for a medium through which funds can be made available to achieve their set goals for society. The government needs money to be able to deliver its social obligations to the public, and these social obligations include, but are not limited to, the provision of infrastructure and social services. Meeting the needs of society calls for huge funds which an individual or society cannot contribute alone and one medium through which funds are derived is through taxation. Tax is a major source of government revenue all over the world. Governments use tax proceeds to provide their traditional functions, such as the provision of public goods, maintenance of law and order, defense against external aggression, regulation of trade and business to ensure social and economic maintenance [3].

Tax morale has been defined as the intrinsic motivation to pay tax and has been linked to "civic duty". Even though tax morale is frequently acknowledged as relevant to tax compliance, little is known about how it comes into being and how it is best nurtured. Loss of legitimacy accompanies less moral obligation to comply, in this case, reduced tax morale. Individuals who feel personally disadvantaged and regard the tax system as responsible for their experiences of hardship are more likely to have depleted levels of tax morale.

A developed economy is one with the ingredients to stimulate investment and create wealth, this, by implication, offers an atmosphere that is business friendly and has the potential for the actualization of Vision 2020. The desired outcome requires a lot of money to put the economy in a position that stimulates investment. Therefore, tax policies need to attract potential investors, and the revenue from tax should be sufficient to meet the infrastructure expenditure of the government [4].

Tax evasion is the illegal concealment of a taxable activity. Measuring how much economic activity is concealed will always be difficult since those who engage in evasion make every effort to hide their activities [5]. Tax compliance refers to the willingness of people to agree with tax authorities by paying their taxes.

Tax compliance indicates the practice of reporting income and paying taxes to a tax administration. The practitioners and academics that have traditionally been most interested in tax compliance, e.g., tax inspectors, accountants and legal scholars, typically describe tax compliance, "as reporting all income and giving all taxes raised by the applicable laws, regulations and court decisions" [6]. Research shows that tax compliance is affected by (social and personal) norms such as those regarding procedural justice, trust, belief in the legitimacy of the government, reciprocity, altruism and identification with the group. Tax non-compliance or evasion, on the other hand, occurs when taxpayers intentionally or unintentionally fail to comply with their tax obligations. Trust-building actions are the most effective economic policies, not only for increasing tax compliance, but also to achieve long-term positive macroeconomic effects.

Voluntary compliance depends on trust in tax authorities, whereas enforced compliance depends on the effectiveness of tax authorities to clamp down on tax evaders. Hence, trust (in) and the power (of) tax authorities are the major determinants for each form of compliance [7].

A self-assessment system (SAS) has become the key administrative approach for both personal and corporate taxation in developed countries including the USA, UK and Australia. This approach emphasizes both the taxpayers' responsibility to declare their income and the need for them to determine their tax liability.

Neoclassical mathematical models of tax behavior conclude that taxpayers will avoid stating their actual revenue to maximize their incomes, an outcome that overpredicts what is observed in reality [8]. According to previous studies on this topic, one of the main facilitating factors in achieving these aims is the development of the level of tax knowledge among taxpayers.

Knowledge transfer is said to take place when one actor in a network is affected by the experience of another and is manifested through any changes in the performance of the recipient of such knowledge [9]. Knowledge has been linked in the literature to terms like data, information, experience, intuition and ideas, depending on the context. Few studies empirically test the connection between knowledge and performance. Thus, it is worth studying how and under which conditions knowledge management (KM) initiatives lead to better results [10].

Nowadays knowledge is an essential resource for modern organizations to support sustainable competitive advantage. Many authors point out that as knowledge is created, disseminated and applied; it contributes to value creation within organizations by enhancing their capabilities to respond to pressures from the external environment. KM has emerged as a scientific discipline to acknowledge the critical role of knowledge in modern organizations. However, the dominant view of KM is technology oriented and considers this activity primarily as an integrated approach to identifying, retrieving, capturing, storing and sharing an organization's information assets. Knowledge is today the most vital resource of modern organizations due to the shift from traditional economic systems based on physical labor to economic systems which rely on knowledge work. Such a transition has been facilitated by the increasing mechanization and automation as well as by information technology which enables organizations to process and make use of ever more information in ever decreasing time cycles. According to the knowledge-based view, knowledge is the central and main strategic resource of organizations. It suggests that knowledge forms the basis for differential organization performance since it transforms individual and social skills and knowhow into economically valuable products and services. The role of organizational actors consists of creating knowledge while the role of organizations is in the integration of individuals' knowledge, and the application of existing knowledge to produce goods and services [11].

Knowledge sharing is vital to knowledge creation and consequent value creation [12]. Knowledge needs a common ground of understanding so that all individuals involved in the knowledge-sharing process can extract whatever is beneficial for them. However, it is necessary to consider that the key to successful knowledge sharing is the availability of the right knowledge at the right time and in the right situation so that an individual group member can do his or her tasks effectively. A focus on the effectiveness of knowledge sharing at the individual and group level is required, taking a dynamic view of knowledge flows.

Preparing information requires extensive effort from experts. Knowledge bases are particularly useful in aiding decision making as expert knowledge can be flexibly captured and utilized. Expert knowledge can be represented as comprehensible rules for decision making in different applications. Another technical barrier is the technical readiness, regarding both the management and the implementation of KM systems. Studies suggest that successful knowledge transfer depends on a host of factors, including mutual learning, an adaptive process, ease of communication, positive source unit – recipient unit link and an organizational culture that fosters knowledge creation and sharing. These factors contribute to the effectiveness of the knowledge transfer process [13]. In attempting to describe knowledge, a broadly accepted distinction refers to tacit knowledge and explicit knowledge. Tacit knowledge cannot be expressed, recounted or articulated in formulas, ratios or drawings: it is knowledge of how to accomplish something (knowhow), as conceptually contrasted with the explicit knowledge of contents ("know that"). It follows that tacit knowledge can be described as "sticky", localized and firmly rooted in the context in which it develops. Explicit knowledge, on the other hand, spreads and is assumed to be "leaky".

While explicit knowledge can be acquired and transferred using rules and norms, tacit knowledge is acquired and transmitted through the sharing of practices, i.e., through the full performance of a task, a job or a profession.

In particular, sharing tacit knowledge about how the knowledge system works is essential to ensure the smooth transfer of technical information. Distance – both geographic and cultural – has been shown to decrease the effectiveness of knowledge transfer. In the last decade, the importance of knowledge has been highlighted by both academics and practitioners. Nowadays, knowledge is the essential basis of competition and, especially tacit knowledge, can be a source of advantage because it is unique, imperfectly mobile, imperfectly imitable and non-substitutable.

In a knowledge-driven economy, knowledge is considered as the economic resource and the only source of competitive advantage. Knowledge sharing is the key to successful KM. Knowledge sharing practices often seem to fail because industry tries to fit KM and knowledge-sharing practices into its existing culture. At present, knowledge sharing is very crucial for industry, but still, individuals do not share their knowledge because they are aware of their value in the organizations. Knowledge sharing can help the individuals to remain valuable in the organizations. A motivational method to encourage a knowledge-sharing attitude but changing the attitude of individuals is one of the biggest challenges for the success of knowledge sharing and KM strategy [14]. The results confirm that dependence and trust maintain a strong impact on knowledge sharing, leading to good team project performance. Findings demonstrate that team members share their knowledge when they trust their partners and when they feel dependent. Empirical evidence suggests that educating taxpayers about the tax system, tax laws and informing them about the negative effects of tax evasion, sanctions and fines is a beneficial policy to increase trust in the authorities, leading to significantly more tax compliance. In addition to tax education, knowledge about tax laws also plays a major role in determining taxpayers' compliance behavior.

Effective knowledge sharing amongst individuals regarding tax would lead to an increase in tax payments. Emphasis and attention to the creation of knowledge in the tax organization would improve its cognition of the environment and perception of the taxpayers' requirements and preferences. As a result, the tax organization could learn new methods and, according to these new procedures, improve the efficiency and effectiveness of the tax system.

In previous studies on tax and related works, the factor of knowledge sharing has not been discussed, but in this chapter, this subject is investigated.

PROPOSED PROBLEM AND MODEL

We consider individuals as taxpayers in the scope of knowledge-sharing as shown in Figure 7.1. These individuals transfer information about paying their taxes with each other. They receive this information from the three categories shown in Figure 7.1. The smallest box is the information about taxes, which consists of an individual's observations, experiences and activities. The second largest box, or the explicit knowledge, is an individual's obvious experiences. In other words, in this box the individual's information about taxes is documented, which consists of information being collected, stored, and archived about the previous year's taxes and the effects

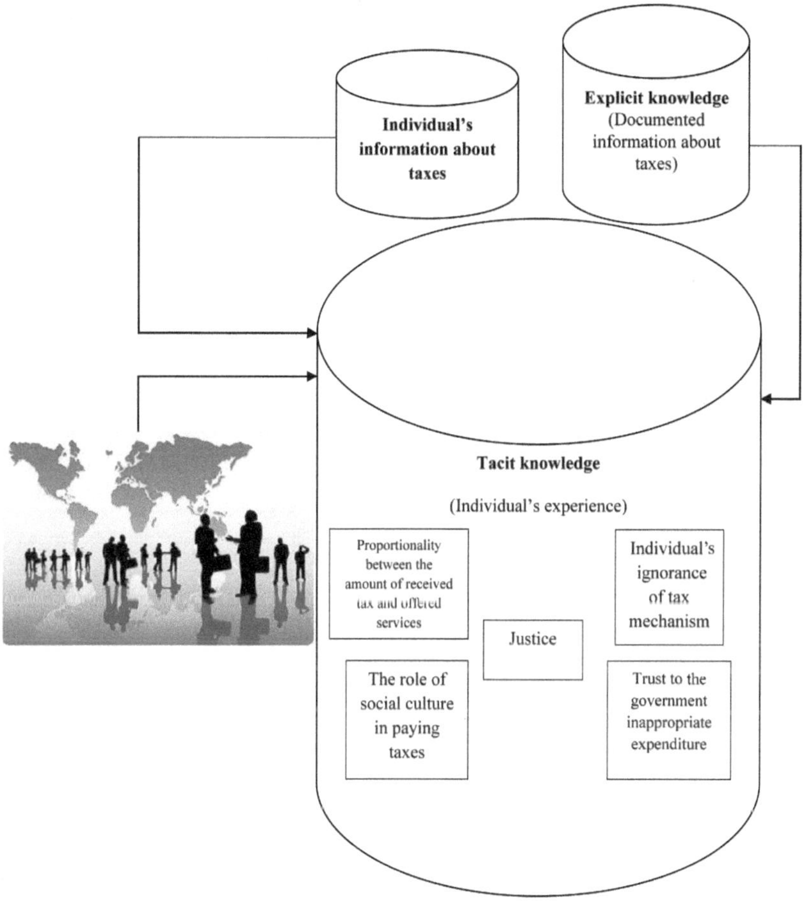

FIGURE 7.1 Tax knowledge sharing system.

of these paid taxes on improvements in facilities, services and so on. The largest box is tacit knowledge or the individual's unobvious experiences about taxes that consist of decreasing unemployment, increasing employment, implementing the law and performing public facilities.

Now the question is how we can fill these boxes? One way is surveying taxpayers and collecting information about the way they share their knowledge.

We have *i* categories (five for the time being) that knowledge is sharing among them. Individuals transfer their knowledge and experiences about these categories with each other and by this means play an important role in tax paying. These categories are collected from surveying academics and studying some articles and economic reports.

1. Justice.

   Unjust tax system, lack of identification of tax resources, inconsistency between tax level and offered services and between the amount of determined tax and type of job.

People believe that taxpayers with higher incomes compared to those with lower incomes must have more responsibility to pay taxes.

2. Proportionality between the amount of tax received and the services offered.

   People believe that there is no correlation between the amount of taxes received and the value of public services offered by the government. In other words, people cannot see the effect of taxes on their lives.

3. Individuals' ignorance of the tax mechanism.

   The government must clarify which projects are funded with taxes paid by the people. Another problem is that these are long-term projects. People do not know why they pay taxes and how these taxes will be used.

4. The role of social culture in paying taxes.

   Not transferring information about taxes, ignorance of the importance of paying tax, non-existence of a culture of taxation, not controlling tax evasion and so on are the most important factors in this category. Some people think that it is government's duty to supply all their needs and because of this, there is no reason for them to pay taxes.

5. Trust in the government to prevent inappropriate expenditure.

   People mostly are distrustful of the government in spending their taxes. At first, the government should spend taxes for public goods and avoid wasting money by spending on luxuries. If the government spends taxes on non-essential items, people cannot trust them and will not make a correct declaration about their income.

Risk Modeling Structure

We introduced five categories earlier of how individuals transfer their knowledge about taxes and related issues in these categories. We apply the information transfer as tax knowledge sharing. Each of these factors causes a degree of risk in receiving taxes. We compute the quality with a risk function. We aim to compute the effectiveness quality of each factor in sharing information [16]. Risk function is an interval function that is defined below:

$$R\left(\underline{x_i}\right) = \int_u^v \underline{L_i} \cdot f\left(x_i\right) d\underline{x_i} \tag{7.1}$$

where $\underline{L_i}$ is loss function and is given by

$$\underline{L_i} = D_{\text{chebyshev}}\left(\underline{t_i}, \underline{x_i}\right) = \left|\underline{t_i} - \underline{x_i}\right| \tag{7.2}$$

and the distribution function *f(x)* shows the behavior of each variable that follows a uniform distribution. The values of x_i are five determined factors and are random variables. We want to compute the risk of x_i values at first and then to compute $\underline{L_i}$. After that, we can compute $\overline{R_i}$. Next, we will introduce interval computations, functions and integrals that will be used in our research.

Interval Programming

Linear programming is among the most widely and successfully used decision tools in the quantitative analysis of practical problems where rational decisions have to be made. The conventional linear programming model requires the parameters to be known as constants. In the real world, however, the parameters are seldom known exactly and have to be estimated. Interval programming is one of the tools to tackle uncertainty in mathematical programming models.

A closed real interval $\left[x_I, x_S\right]$ denoted by $\underline{x}$, is a real interval number which can be defined completely by

$$\underline{x} = \left[x_I, x_S\right] = \left\{x \in R \,\middle|\, x_I \le x \le x_S; x_I, x_S \in R\right\}$$

where x_I and x_S are called infimum and supremum, respectively.

Two interval numbers $x = [x_I, x_S]$ and $y = [y_I, y_S]$ are called equal if and only if $x_I = y_I$ and $x_S = y_S$ [15].

Let $\underline{x} = \left[x_I, x_S\right]$ and $\underline{y} = \left[y_I, y_S\right]$, then

1. $\underline{x} + \underline{y} = \left[x_I + y_I, x_S + y_S\right]$ (Addition)
2. $\underline{x} - \underline{y} = \left[x_I - y_S, x_S + y_I\right]$ (Subtraction)
3. $\underline{xy} = \left[\min\left\{x_I y_I, x_I y_S, x_S y_I, x_S y_S\right\}, \max\left\{x_I y_I, x_I y_S, x_S y_I, x_S y_S\right\}\right]$ (Multiplication)

   Let

$$\underline{x}, \underline{y} \,\&\, \underline{z} \in I(R)$$

5. $\underline{x} + \underline{y} = \underline{y} + \underline{x}$, $\underline{xy} = \underline{yx}$ (Commutativity)
6. $\left(\underline{x} + \underline{y}\right) + \underline{z} = \underline{x} + \left(\underline{y} + \underline{z}\right), \left(\underline{xy}\right)\underline{z} = \underline{x}\left(\underline{yz}\right)$ (Associativity)
7. $\underline{x}\left(\underline{y} + \underline{z}\right) \subseteq \left(\underline{xy} + \underline{xz}\right)$ (Subdistributivity)
8. $a\left(\underline{x} + \underline{y}\right) = a\underline{x} + a\underline{y}, a \subseteq R.$

The construction of the interval integral, given in the general case, can be simplified drastically in the case that the interval of integration is finite and the integrand is a bounded interval function. (Definitions of the necessary concepts will be given below.) In particular, the use of the extended real number system is not required, so that all computations can be done by ordinary interval arithmetic.

Following the definitions, an interval function Y defined on an interval $X = [a,b]$ assigns the interval value,

$$Y(x) = \left[\underline{y}(x), \overline{y}(x)\right] \tag{7.3}$$

to each real number $x \in X$, where $\underline{y}$, $\overline{y}$ are real functions called, respectively, the lower and upper boundary functions (or endpoint functions) of Y.

In general, the interval integral of an interval function Y over the interval $X = [a,b]$ is the interval,

$$\int Y(x)\mathrm{d}x = \int_a^b Y(x)\mathrm{d}x = \left[\underline{\int} \underline{y}(x)\mathrm{d}x, \overline{\int} \overline{y}(x)\mathrm{d}x\right] \tag{7.4}$$

where $\underline{\int} \underline{y}(x)\mathrm{d}x$ denotes the lower Darboux integral of the lower endpoint function $\underline{y}$ over the interval X and $\overline{\int} \overline{y}(x)\mathrm{d}x$ gives the upper Darboux integral of the upper endpoint function $\overline{y}$ over X. As these Darboux integrals always exist in the extended real number system, it follows that all interval (and hence all real) functions are integrable in this sense.

We consider the below algorithm for the explained parts:

1. Identifying the effective factors in tax paying.
   We mean collected categories from surveying academics and studying some articles and economic reports that are explained above.
2. Effective factors in tax paying are estimated by knowledge sharing.
3. Computing the risk function of each factor in tax knowledge sharing.

A risk is the effect of each factor in tax knowledge sharing. In this part, we will use the presented formulas for computing risk. The reason for using the risk function here is the uncertainty in sharing taxpayers' opinions. This uncertainty is caused by the fact that a factor could be effective as α in a case study while as β in another one. It is also possible to have both opponents and proponents about a factor. As a result, there is no definitive opinion about knowledge-sharing factors.

Now we draw a flowchart for the above algorithm (see Figure 7.2).

CASE STUDY

As stated before, we considered five categories of ways in which individuals share their tax knowledge. These categories were collected from surveying academics and studying some articles and economic reports. Briefly, the categories are justice, the proportionality between the amount of received tax and services offered, people's ignorance of the tax mechanism, the role of social culture in paying taxes and trust in the government to prevent inappropriate expenditure. To gather the results of knowledge sharing between individuals we decided to survey the taxpayers. So a survey form was prepared (see Figure 7.3).

We asked 65 academics and experts to complete the survey form. The results of this survey are collected in Table 7.1. In Table 7.1, the first factor expresses justice in the tax system and shows the individual's ideas about justice, that one person has chosen option one, five persons have chosen option two, 18 persons agree with option four and one person has chosen option five. Respectively, the second factor expresses proportionality between the amount of tax received and services offered by the tax system; the third one expresses individuals' knowledge about the tax mechanism; the fourth factor is the role of social culture in paying taxes and finally the last one

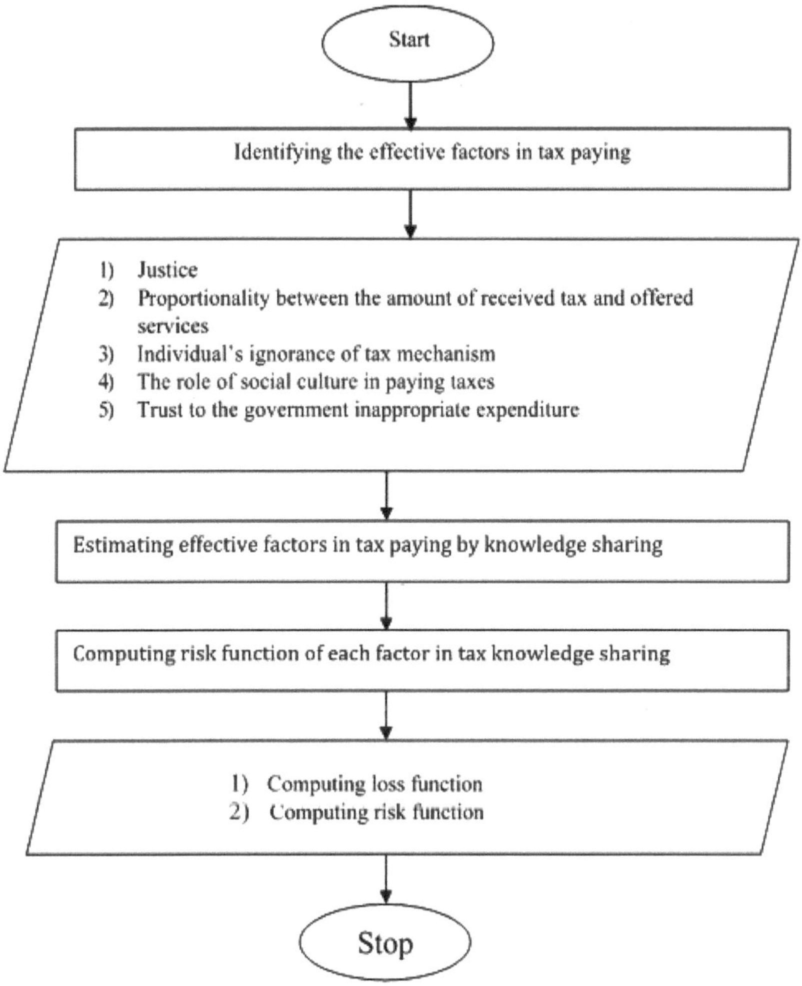

FIGURE 7.2 An algorithm.

expresses the role of trust in the government to prevent inappropriate expenditure the of taxes received.

Computing Mean Values (t_i)

t_i is the mean value of the achieved response for each factor. We compute t_i for the factors. The solution method, in this case, is that we equal interval (0,0.25) to the first option, interval (0.25,0.5) to the second option, interval (0.5,0.75) to the third option and interval (0.75,1) to the last option. Then we multiply the number of individuals who had chosen the first option by interval (0.0.25) and multiply the number of individuals who had chosen the second option by interval (0.25,0.5) and so on to intervals (0.5,0.75) and (0.75,1). Next, we sum the above four achieved intervals and

1) How would you rate justice by tax authorities?

   1) very high ☐ 2) high ☐ 3) moderate ☐ 4) low ☐

2) How would you rate proportionality between the amount of received tax and offered services in the tax system?

   1) very high ☐ 2) high ☐ 3) moderate ☐ 4) low ☐

3) How would you rate individual's knowledge of tax mechanism?

   1) very high ☐ 2) high ☐ 3) moderate ☐ 4) low ☐

4) How would you rate the role of social culture in paying taxes?

   1) very high ☐ 2) high ☐ 3) moderate ☐ 4) low ☐

5) How would you rate the role of trust to the government inappropriate expenditure of received taxes?

   1) very high ☐ 2) high ☐ 3) moderate ☐ 4) low ☐

FIGURE 7.3 Survey form.

then divide this summation by 65, that is the number of individuals who took part in the survey. After receiving qualitative responses, for applying to formulas we use them in the normal distance (*N(0,1)*). We divided this distance by four because the questions consist of four options. It means that the first option "very high", is equal to (0,0.25), the second option "high", is equal to (0.25,0.5), and respectively options three and four are equal to (0.5,0.75) and (0.75,1).

$$\underline{t_1} = \frac{1*(0,0/25)+5*(0/25,0/5)+18*(0/5,0/75)+41*(0/75,1)}{65}$$

$$= \frac{(0,0/25)+(1/25,2/5)+(9,13/5)+(30/75,41)}{65}$$

$$= \frac{(41,57/25)}{65} = (0/63,0/88)$$

TABLE 7.1
The Results of the Survey

	Option One	Option Two	Option Three	Option Four
First factor	1	5	18	41
Second factor	0	8	20	37
Third factor	1	2	18	44
Fourth factor	11	12	15	27
Fifth factor	0	7	21	37

$$\underline{t}_2 = \frac{0*(0,0/25)+8*(0/25,0/5)+20*(0/5,0/75)+37*(0/75,1)}{65}$$

$$= \frac{(0,0)+(2,4)+(10,15)+(27/75,37)}{65}$$

$$= \frac{(39/75,56)}{65} = (0/61,0/86)$$

$$\underline{t}_3 = \frac{1*(0,0/25)+2*(0/25,0/5)+18*(0/5,0/75)+44*(0/75,1)}{65}$$

$$= \frac{(0,0/25)+(0/5,1)+(9,13/5)+(33,44)}{65}$$

$$= \frac{(42/5,58/75)}{65} = (0/65,0/90)$$

$$\underline{t}_4 = \frac{11*(0,0/25)+12*(0/25,0/5)+15*(0/5,0/75)+27*(0/75,1)}{65}$$

$$= \frac{(0,2/75)+(3,6)+(7/5,11/25)+(20/25,27)}{65}$$

$$= \frac{(30/75,47)}{65} = (0/47,0/72)$$

$$\underline{t}_5 = \frac{0*(0,0/25)+7*(0/25,0/5)+21*(0/5,0/75)+37*(0/75,1)}{65}$$

$$= \frac{(0,0)+(1/75,3/5)+(10/5,15/75)+(27/75,37)}{65}$$

$$= \frac{(40,56/25)}{65} = (0/61,0/86)$$

COMPUTING LOSS FUNCTION $(\underline{L}_i)$

In this section, by means of $\underline{t}_i$ values that were computed in the previous part we can compute loss functions. We use the loss formula that was defined in the 'Risk Modeling Structure' section for computing loss function for all factors:

$$\underline{L}_i = D_{\text{chebyshev}}(\underline{t}_i, \underline{x}_i) = |\underline{t}_i - \underline{x}_i|$$

$\underline{x}_i$ values are variable. The computed values for $\underline{t}_i$ are as below:

$$\underline{t}_1 = (0.63, 0.88)$$

$$\underline{t_2} = (0.61, 0.86)$$

$$\underline{t_3} = (0.65, 0.90)$$

$$\underline{t_4} = (0.47, 0.72)$$

$$\underline{t_5} = (0.61, 0.86)$$

Now we compute $\underline{L_i}$ functions by putting these intervals in a loss function formula:

$$\underline{L_1} = D_{\text{chebyshev}}\left(\underline{t_1}, \underline{x_1}\right) = \left|\underline{t_1} - \underline{x_1}\right| = \left|\left(t_{1I}, t_{1S}\right) - \left(x_{1I}, x_{1S}\right)\right|$$

$$= \left|\left(0.63, 0.88\right) - \left(x_{1I}, x_{1S}\right)\right| = \left(0.63 - x_{1S}, 0.88 + x_{1I}\right)$$

$$\underline{L_2} = D_{\text{chebyshev}}\left(\underline{t_2}, \underline{x_2}\right) = \left|\underline{t_2} - \underline{x_2}\right|$$

$$= \left|\left(t_{2I}, t_{2S}\right) - \left(x_{2I}, x_{2S}\right)\right| = \left|\left(0.61, 0.86\right) - \left(x_{2I}, x_{2S}\right)\right|$$

$$= \left(0.61 - x_{2S}, 0.86 + x_{2I}\right)$$

$$\underline{L_3} = D_{\text{chebyshev}}\left(\underline{t_3}, \underline{x_3}\right) = \left|\underline{t_3} - \underline{x_3}\right|$$

$$= \left|\left(t_{3I}, t_{3S}\right) - \left(x_{3I}, x_{3S}\right)\right| = \left|\left(0.65, 0.90\right) - \left(x_{3I}, x_{3S}\right)\right|$$

$$= \left(0.65 - x_{3S}, 0.90 + x_{3I}\right)$$

$$\underline{L_4} = D_{\text{chebyshev}}\left(\underline{t_4}, \underline{x_4}\right) = \left|\underline{t_4} - \underline{x_4}\right|$$

$$= \left|\left(t_{4I}, t_{4S}\right) - \left(x_{4I}, x_{4S}\right)\right| = \left|\left(0.47, 0.72\right) - \left(x_{4I}, x_{4S}\right)\right|$$

$$= \left(0.47 - x_{4S}, 0.72 + x_{4I}\right)$$

$$\underline{L_5} = D_{\text{chebyshev}}\left(\underline{t_5}, \underline{x_5}\right) = \left|\underline{t_5} - \underline{x_5}\right|$$

$$= \left|\left(t_{5I}, t_{5S}\right) - \left(x_{5I}, x_{5S}\right)\right| = \left|\left(0.61, 0.86\right) - \left(x_{5I}, x_{5S}\right)\right|$$

$$= \left(0.61 - x_{5S}, 0.86 + x_{5I}\right)$$

Computing $f(x_i)$ for the Factors

$f(x_i)$ follows a uniform distribution. Random variables x_i is said to distribute uniformly in the interval $[a,b]$ if it has the probability density function as below:

$$f(x_i) = \begin{cases} \dfrac{1}{b-a} & a \le x_i \le b \\ 0 & \text{otherwise} \end{cases}$$

The support is defined by the two parameters, *a* and *b*, which are its minimum and maximum values, respectively.

Here *a* and *b* will be achieved by the survey. For each factor, *b* is obtained by multiplication of the second component of attributed interval (0.75,1) to the fourth option that is number 1 with the number of individuals who had chosen the fourth option for each factor. Also, *a* is obtained by multiplication of the second component of attributed interval (0,0.25) to the first option, that is number 0.25, with the number of individuals who had chosen the first option for each factor.

According to Table 7.1, for the factors we compute *a* and *b* and after that $f(x_i)$ for each factor:

For x_1:

$$b = 1 * 41 = 41$$

$$a = 0.25 * 1 = 0.25$$

$$f(x_1) = \frac{1}{b-a} = \frac{1}{41-0.25} = \frac{1}{40.75} = 0.019$$

For x_2:

$$b = 1 * 37 = 37$$

$$a = 0.25 * 0 = 0$$

$$f(x_2) = \frac{1}{b-a} = \frac{1}{37-0} = \frac{1}{37} = 0.027$$

For x_3:

$$b = 1 * 44 = 44$$

$$a = 0.25 * 1 = 0.25$$

$$f(x_3) = \frac{1}{b-a} = \frac{1}{44-0.25} = \frac{1}{33.75} = 0.03$$

For x_4:

$$b = 1 * 27 = 27$$

$$a = 0.25 * 11 = 2.75$$

$$f(x_4) = \frac{1}{b-a} = \frac{1}{27-2.75} = \frac{1}{24.25} = 0.041$$

For x_5:

$$b = 1 * 37 = 37$$

$$a = 0 / 25 * 0 = 0$$

$$f(x_5) = \frac{1}{b-a} = \frac{1}{37-0} = \frac{1}{37} = 0.027$$

Computing Risk Function ($R(\underline{x}_i)$)

All of the five factors create a different degree of risk in tax paying. We have to compute the risk function to control the effectiveness quality of each factor in tax knowledge sharing among individuals.

As stated before for computing risk we use the formula below:

$$R(\underline{x_i}) = \int_u^v \underline{L_i} . f(x_i) d\underline{x_i}$$

$\underline{L_i}$ is loss function that was computed before for the factors, also $f(x_i)$ has been introduced before.

For integral bounds, we do as follows:

For each factor, we consider the first option that has a non-zero answer (it means that at least one person has chosen the option), then, the supremum value of the attributed interval to this option is chosen as the lower bound (u), and the next non-zero answer for the same factor is chosen as the supremum value of the attributed interval to this option as the upper bound (v). In the following computations the upper bound and lower bound for each factor will compute. Then, using loss functions and $f(x_i)$ we will compute risk function for each factor.

First factor $(\underline{x_1})$:

Integral bounds:

We have no zero answers among the answers for the first factor, so we consider the interval (0,0.25) attributed to the first option. The supremum value of this interval is 0.25. So, we choose $u = 0.25$ as the lower bound. The attributed interval to the fourth option is (0.75,1) that the supremum value of it is 1, so we choose $v = 1$ as the upper bound.

Also, we computed $\underline{L_1}$ and $f(x_1)$ as below:

$$\underline{L_1} = (0.63 - x_{1S}, 0.88 + x_{1l})$$

$$f(x_1) = 0.019.$$

So, we compute $R(\underline{x_1})$ as below:

$$R(\underline{x_1}) = \int_{0/25}^{1} \underline{L_1} \cdot f(x_1) \cdot d\underline{x_1}$$

$$= \int_{0/25}^{1} \underline{L_1} \cdot 0/019 \cdot d\underline{x_1}$$

$$= 0/019 * \int_{0/25}^{1} \underline{L_1} \cdot d\underline{x_1}$$

$$= 0/019 * \int_{0/25}^{1} (0/63 - x_{1S}, 0/88 + x_{1I}) \cdot d\underline{x_1}$$

$$= 0/019 * \left[\int_{0/25}^{1} (0/63 - x_{1S}) \cdot dx_{1S}, \int_{0/25}^{1} (0/88 + x_{1I}) \cdot d\underline{x_{1I}} \right]$$

$$= 0/019 * \left[\left(0/63x_{1S} - \frac{1}{2} x_{1S}^2 \right), \left(0/88x_{1I} + \frac{1}{2} x_{1I}^2 \right) \right]_{0/25}^{1}$$

$$= 0/019 * [(0/63 - 0/1575) - (0/5 - 0/0312),$$

$$(0/88 - 0/22) + (0/5 - 0/0312)]$$

$$= 0/019 * [0/0037, 1/1288] = [0/00007, 0/02145]$$

TABLE 7.2
A Summary of Risk Computations

	$\underline{L_i}$	$f(x_i)$	$\underline{R_i}$
Second factor	$\underline{L_2} = (0.61 - x_{2S}, 0.86 + x_{2I})$	$f(x_2) = 0.027$	$R(\underline{x_2}) = [0.07, 0.022]$
Third factor	$\underline{L_3} = (0.65 - x_{3S}, 0.90 + x_{3I})$	$f(x_3) = 0.03$	$R(\underline{x_3}) = [0.000561, 0.0343]$
Fourth factor	$\underline{L_4} = (0.47 - x_{4S}, 0.72 + x_{4I})$	$f(x_4) = 0.041$	$R(\underline{x_4}) = [-0.00477, 0.04136080]$
Fifth factor	$\underline{L_5} = (0.61 - x_{5S}, 0.86 + x_{5I})$	$f(x_5) = 0.027$	$R(\underline{x_5}) = [-0.00189, 0.0217350]$

Then we compute the other factors as above. The results of computations are in Table 7.2.

CONCLUDING REMARKS

In recent decades, tax policy has been at the center of public debate. The economic, social and cultural effects of tax payment is one of the most important topics that needs extra attention from every government. By focusing on the role of tax knowledge sharing in tax payments that have not been mentioned previously in other studies, this chapter adds knowledge sharing as an important factor in tax systems. Since taxation is an instrument of economic growth, policies to improve a tax system's productivity and efficiency should be implemented and sustained. This conclusion points to the need for additional attempts by governments in educating individuals in tax knowledge and encouraging them to share and transfer that knowledge with others to increase tax compliance by taxpayers, and as a result to increase tax revenues for better facilities and for growth.

Here, we introduced five categories in which taxpayers share their knowledge. We considered individuals as taxpayers in knowledge sharing. They transfer their knowledge and experiences about these categories with each other and play an important role in tax raising. These categories are collected from a survey which was extracted from studying some articles and economic reports. We apply this transferring of information as tax knowledge sharing. Each of these factors caused a degree of risk in receiving taxes. We computed risk for each factor using a loss function.

REFERENCES

1. Lymer A, Oats L. *Taxation: policy and practice.* 16th ed. Birmingham: Fiscal Publications; 2009.
2. Palil MR. *Tax knowledge and tax compliance determinants in self-assessment system in Malaysia* (Doctoral dissertation). University of Birmingham, UK; 2010.
3. Bukie HO, Adejumo TO. The effects of tax revenue on economic growth in Nigeria (1970–2011). *I J Hum Soc Sci Inven.* 2013;2(6):16–26.
4. Ihenyen CJ, Mieseigha EG. Taxation as an instrument of economic growth (the Nigerian perspective) *Inf Knowl Manag,* 2014;4(12):49–55.
5. Hashimzade N, Myles GD, Page F, Rablen MD. Social networks and occupational choice: The endogenous formation of attitudes and beliefs about tax compliance. *J Econ Psy.* 2014;40:134–46.
6. Boll K. Mapping tax compliance Assemblages, distributed action and practices: A new way of doing tax research. *Crit Pers Account.* 2013;1:25–42.
7. Lisi G. The interaction between trust and power: Effects on tax compliance and macroeconomic implications. *J Behav Experim Econ.* 2014;53:24–33.
8. Andrei AL, Comer K, Koehler M. An agent-based model of network effects on tax compliance and evasion. *J Econ Psych.* 2014;40:119–33.
9. Tagliaventi MR, Bertolotti F, Macri DM. A perspective on practice in interunit knowledge sharing. *Eur Manag J.* 2010;28(5):331–45.
10. Lopez-Nicolas, C, Meroño-Cerdan AL. Strategic knowledge management, innovation and performance. *Int J Inf Manag.* 2011;31:502–9.
11. Guetat SA., Dakhli SD. A framework for understanding the complementary roles of information systems and knowledge management systems. *Int J Knowl Org.* 2014;4(3):39–55.
12. Zhang X, Mao X., AbouRizk SM. Developing a knowledge management system for improved value engineering practices in the construction industry. *Aut Const.* 2009;18:777–89.
13. Ngoma NS, Lind M. Knowledge transfer and team performance in distributed organizations. *Int J Knowl Org.* 2015;5(2):58–80.

14. Sharma BP, Singh MD. Modeling the knowledge sharing barriers: An ISM approach. *Int J Knowl Org*. 2015;5(1):16–33.
15. Suprajitno H, Mohd I. Linear programming with interval arithmetic. *Int J Contemp Math Sci*. 2010;7(5):323–32.
16. Kazemi Z., Fazlollahtabar H. Integrated model of knowledge sharing and tax willingness to pay. *Int J Oper Quant Manag*. 2017;23(4):295–315.

8 Knowledge Sharing for Enterprise Resources

INTRODUCTION AND BACKGROUND

Knowledge is a substantial factor in organizations, providing a sustainable competitive advantage in a dynamic market environment. To obtain a competitive advantage it is crucial but insufficient for organizations to place emphasis on staffing and training systems that concentrate on employees' selection based on specific knowledge, skills, abilities or competencies or helping employees acquire them. Organizations should also consider how to transfer expertise and knowledge from a person to others who need to know. The concept of knowledge management (KM) has attracted the attention of researchers over the last decade since it is considered an important tool to achieve innovation and sustainable competitive advantage.

Extensive research has demonstrated the importance of customer-employee interactions in customers' evaluation of overall quality and/or satisfaction with services [1]. Knowledge sharing among employees and within and across enterprise resources leads organizations to exploit and utilize their knowledge-based resources. There is substantial evidence of the impact of KM practices in building strong relationships with customers, and enhancing customer satisfaction and organizational performance [2–4].

However, no prior studies have investigated the influence of KM practices in a service encounter context. KM begins with an understanding that knowledge is broadly classified as explicit knowledge and tacit knowledge. Each has very particular characteristics that influence KM, discussed more fully later in this chapter. An important reason for the failure of the knowledge management system (KMS) to facilitate knowledge sharing is the lack of attention of how the organizational and interpersonal context, as well as individual characteristics, affects knowledge sharing.

The framework shown in Figure 8.1 was based on the review of the literature and presents a structure for the proposed work. Figure 8.1 illustrates five areas of emphasis in knowledge sharing research, the topics within each area of emphasis that have been explored, and the relationships between each area of emphasis and knowledge sharing. As shown in Figure 8.1, the topics studied within each area of emphasis have been shown to influence knowledge sharing directly or indirectly through motivational factors. The right side of Figure 8.1 indicates the common dependent variables surveyed in the literature (knowledge sharing intention, intention to encourage knowledge sharing and knowledge sharing behaviors). Also, in Figure 8.1, the topics shown in the shaded boxes with solid lines have been examined in the existing literature. The topics shown in the boxes with dotted lines need future research. The topics shown in the overlapping areas represent those that have been examined in prior

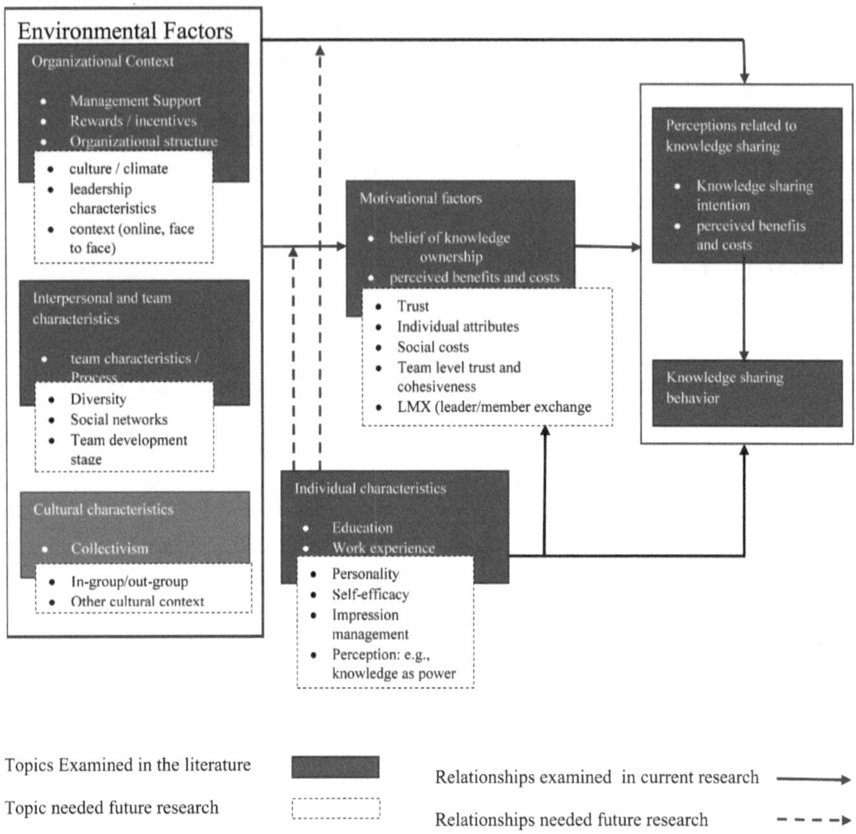

FIGURE 8.1 A framework of knowledge sharing.

studies but still warrant further research. Third, this review contributes to human resource management practice by discussing the implications of knowledge sharing research for the implementation, support and effectiveness of knowledge sharing initiatives in organizations.

There is a need to deepen the inquiry into the emergent and rapid development of KM in China and Chinese contexts. The extant literature on KM is largely either developed theoretically, or based on Western and Japanese business environments from the 1990s through to the present (cf. [5]). While source trustworthiness leads to enhanced knowledge transfer across units, the perception of an individual being trusted by the recipient may also affect his/her motivation to share knowledge with this person. Also, conditional and unconditional trust may have different relationships with knowledge sharing.

A number of researchers have questioned whether researchers are able to understand what happens in other areas because of the cultural, historical and economic differences. Management tools applied elsewhere to knowledge sharing may not be applicable or viable when studying many managerial and other issues in the context. The ideas, experiences and knowledge shared in a KMS are considered to be public goods which are accessible to every member of the system and their value will not be ignored. Several studies included in the review used social capital and network theories.

Knowledge sharing (KS) has become more widely recognized as a key KM construct, especially in service sectors such as consulting. The importance of effective knowledge sharing and resultant innovation from cross pollination of ideas leading to increased competitiveness in Chinese markets has led to Chinese practitioners and organizations seeking to contextualize or to increase their understanding of KS in a Chinese setting. One of the key issues is to understand to what extent the cultural factors have impacted on KS and resultant innovation activities within. When a formal or informal group (or community of practice) is formed, its members bring with them not only their own knowledge, skills and abilities but also their social connections. Based on structural holes, the more employees are bridging structural holes the more likely different types of knowledge may be shared. Therefore, it would be interesting to investigate how employees' network positions are related to knowledge sharing and how organizations may better leverage individuals in these critical positions (for example, see [6]).

The influence of attitudes toward KS on KS intentions and behavior has been investigated rather extensively using the theory of reasoned action. However, few studies have examined their antecedents. For example, [7] showed that the richness of channels for KS and one's absorptive capability to learn from others has a positive influence on individuals' attitudes toward KS. They argued that individuals with higher absorptive capacity are more likely to experience the benefits of KS resulting in more positive attitudes towards it. Future research will benefit from focusing on understanding how to enhance positive attitudes toward KS.

Furthermore, although the role of motivation has been recognized and emphasized in the KS literature, it is somewhat surprising that traditional motivation theories, such as expectancy theory and social cognitive theory have not been used as often in KS research. Future research should investigate KS using these theoretical frameworks, given the insight these theories have provided in understanding other types of voluntary employee behavior such as participation in training.

It should be mentioned that KS in social networks is also an important issue that has been recently studied [8–10]. In some organizations employees consider KS an extra-role behavior, i.e., it is not included in formal job descriptions, while in others it is considered an in-role behavior because KS is expected and is evaluated and/or rewarded. Future research needs to investigate whether there are differences in the type or quality of knowledge shared when it is considered an in-role versus extra-role behavior. Theories related to prosocial organizational behavior and personality may be useful for increasing our understanding of KS when it is considered an extra-role behavior.

The main work of this chapter is to propose a switch architecture-enabling KS paradigm amongst the resources of an enterprise. The proposed switch is associated with three major measurement factors based on mathematical Boolean scales.

PROPOSED FRAMEWORK

We consider a switching framework to process KS activities in an enterprise based on two measures of interest and link. Our objectives to propose a KS process in an enterprise are listed below:

- Determining important enterprise resources after knowledge sharing;
- Configuring a 0/1 switch on the enterprise resources;
- Strengthening knowledge sharing;
- Determining bottlenecks for knowledge sharing.

Also, the following assumptions are considered in the modeling process:

- Enterprise resources are available;
- Information flows exist among enterprise resources;
- KS let to value added;
- KS between enterprise resources;
- Determining cross points in KS and designing switch.

The concept of switch implies determining the knowledge sharing points. In the proposed model, the enterprise resources are categorized based on the organizational chart to have necessary link or unnecessary link where the aim is to fortify the parts having necessary links [11]. We make use of zero/one values in an enterprise function matrix (F), so that the value in the matrix is 1 when the link is necessary and it is zero otherwise. A hypothetical function matrix having the values is given below (note that the elements of the matrix are the resources of the enterprise):

$$F = \begin{bmatrix} 1 & 1 & 0 & 0 & 0 & 0 \\ 1 & 1 & 0 & 0 & 0 & 0 \\ 0 & 0 & 1 & 1 & 1 & 0 \\ 0 & 0 & 1 & 1 & 1 & 1 \\ 0 & 0 & 1 & 1 & 1 & 0 \\ 0 & 0 & 0 & 1 & 0 & 1 \end{bmatrix}$$

Also, the concept of interest between any pair of the human resources in the enterprise is another aspect in the model. Thus, if the two human resources have interest, the value 1 is recorded in the interest matrix (I) and otherwise the value is zero. A hypothetical interest matrix having the values is given below (note that the elements of the matrix are the resources of the enterprise):

$$I = \begin{bmatrix} 1 & 0 & 1 & 1 & 1 & 0 \\ 0 & 1 & 1 & 1 & 0 & 0 \\ 1 & 1 & 1 & 0 & 0 & 1 \\ 1 & 1 & 0 & 1 & 1 & 1 \\ 1 & 0 & 0 & 1 & 1 & 1 \\ 0 & 0 & 1 & 1 & 1 & 1 \end{bmatrix}$$

Now, the cross points are obtained by vector multiplication of two matrices I and F.

$$F \times I = \begin{bmatrix} 1 & 0 & 0 & 0 & 0 & 0 \\ 0 & 1 & 0 & 0 & 0 & 0 \\ 0 & 0 & 1 & 0 & 0 & 0 \\ 0 & 0 & 0 & 1 & 1 & 1 \\ 0 & 0 & 0 & 1 & 1 & 0 \\ 0 & 0 & 0 & 1 & 0 & 1 \end{bmatrix}$$

The values of one in the above matrix show the points that opted for knowledge sharing and should be fortified.

The enterprise resources are configured and the problem-solving process begins here. The essence of an assessment program to evaluate, integrate, deposit and guide enterprise resources achievements during the process is of importance. The structure of the interactivity network influences the aforementioned assessment program. In spite of the growing consensus that networks matter, the specific effects of different elements of network structure on KS remain widely unclear. In the network literature, a debate has arisen over the network structures that can appropriately be regarded as beneficial. According to one view, close networks with many strong connections linking enterprise resources are seen as advantageous. The alternative view, however, states that advantages derive from the opportunities created by an open social structure. Enterprise resources can build contacts with multiple disconnected clusters of enterprise resources and use these connections to obtain the right information at the right time. From a theoretical point of view, these arguments have different, even contradictory, implications. The closeness of an enterprise resources network is described by its "network range" and "tie strength". Openness is captured by "network efficiency". The knowledge development program (KDP) of our study is shown in Figure 8.2. In this KDP, a performance assessment stage is considered. The aim of this performance assessment is to track the enterprise resources' improvements in the problem-solving process. We conduct a comprehensive network interaction assessment method (CNIAM) for evaluating the performance of the enterprise resource's network interactions in intelligent information exchange. The CNIAM consists of three elements: network range, tie strength and network efficiency [12].

There are numerous other characteristics of enterprise resources that may have been shown to influence KDP enterprise resource KS. Two variables affecting the KS are the research and development (R&D) intensity and the level of innovation. Since both factors are the same for all the enterprise resources being considered, all enterprise resources face a very high R&D intensity and a very high level of innovation, and thus we could not include these factors. Drawing on the innovation literature, we included three control variables. These variables are the size and the experience of the enterprise resources, and the presence of members of various nationalities. Concerning these variables, there are clear variations among the enterprise resources in the KDP.

Network Range

An enterprise resources network range represents the degree to which an enterprise resource is linked to other enterprise resources. Network range has long been

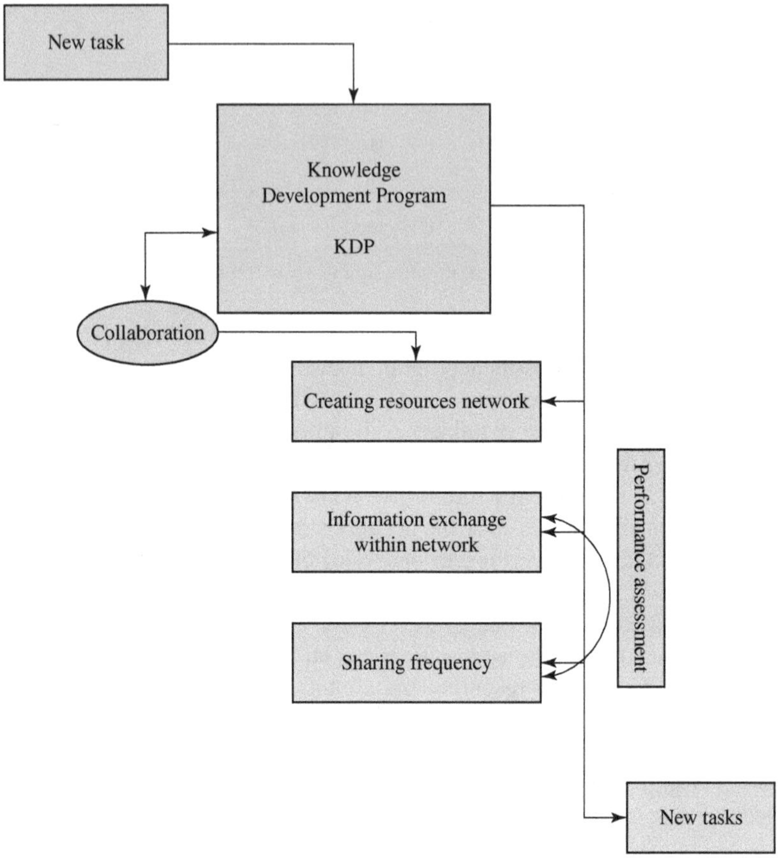

FIGURE 8.2 Knowledge development program for knowledge sharing matrix.

shown to influence how information spreads among enterprise resources in projects. Timely provision with the right information is one of the most important ingredients for knowledge sharing. Having many contacts should thus increase the chances to obtain the required information at the right time by having multiple direct sources of information. Because of its obvious advantages and importance, network range can be expected to positively impact the enterprise resources' KS in KDPs.

Tie Strength

The strength (frequency) of contacts among enterprise resources can affect how easily information is transferred. Enterprise resources that interact frequently with other enterprise resources are more likely to share information than those who interact infrequently. More frequent interaction, especially among those with different expertise, can lead to more effective interaction since the mutual understanding improves. Also, better mutual understanding eases the distribution of sophisticated and complex information. To be able to understand and digest information, engineers need to know the context of the information. This context is learned through intense interaction with others. Also, frequent contacts would also facilitate the formation of trust, which may further ease the transfer of information. Trust gives parties the confidence that the information shared will not be appropriated or misused. Based on this discussion, it can be expected that workgroups with more frequent interaction are more creative in the process of innovation.

Network Efficiency

The third aspect of enterprise resource network considered addresses the degree of connectivity and interactions (or the lack of them) among the enterprise resources. Contacts in interaction networks are redundant to the degree that they lead to the same enterprise resources. An efficient network structure indicates that enterprise resources have access to different non-overlapping flows of information. Efficient networks imply access to mutually unconnected partners and, consequently, guarantee enterprise resources more autonomy, while providing sufficient information. Such autonomy has been argued as instrumental to creative achievements. In addition, an efficient network yields less waste of time and energy, and so more can be employed in transforming available knowledge into new knowledge. Using these elements, the proposed CNIAM is configured to analyze the network interactions. These interactions are related to the information sharing among the enterprise resources. Enterprise resources interact and cooperate to solve problems requiring the available knowledge of other enterprise resources. Information sharing takes place during interactions among enterprise resources. To track and record these interactions, we configure an intelligent information network.

Measuring CNIAM

We make use of the following entities:

Network range: Network range is measured by the fraction of all contacts each enterprise resource maintains with the other enterprise resources at least with a monthly frequency.

Tie strength: Tie strength represents the proportional strength of contacts an enterprise resource maintains on the scales of 0 = no interaction, 1 = (at least) monthly interaction, 2 = (at least) weekly interaction, and 3 = (at least) daily interaction.

Network efficiency: To calculate the efficiency of the enterprise resources network we use the efficiency measure. Network efficiency is calculated as the proportion of an enterprise resource's non-redundant relationships.

The measures employed in this work are summarized in Table 8.1.

CASE STUDY

Here, to verify our proposed KS matrix methodology, we consider a case study at an information technology services company. Here, we first collect the KS parameters

TABLE 8.1
The CNIAM Measures

Network Range $NR(n_i)$ = percentage of contacts of item i to all other enterprise resources j N = number of enterprise resources $x_{ij} = \begin{cases} 1 & \text{if } i \text{ is connected to } j \\ 0 & \text{if } i \text{ is not connected to } j \end{cases}$	$NR(n_i) = \frac{\left(\sum_{\forall j \neq i} x_{ij}\right) \times 100}{N}$
Tie Strength $TS(n_i)$ = proportional tie strength of enterprise resource i to all other contacts j $S_{\max}$ = maximum tie strength 3 $x_{ij} = \begin{cases} 1 & \text{if } i \text{ is connected to } j \\ 0 & \text{if } i \text{ is not connected to } j \end{cases}$ $s_{ij} = \begin{cases} 1 & \text{if } i \text{ is connected to } j \text{ at least monthly} \\ 2 & \text{if } i \text{ is connected to } j \text{ at least weekly} \\ 3 & \text{if } i \text{ is connected to } j \text{ at least daily} \end{cases}$	$TS(n_i) = \frac{\left(\sum_{\forall j \neq i} x_{ij} \times s_{ij}\right) \times 100}{\sum_{\forall j \neq i} x_{ij} \times s_{\max}}$
Network Efficiency $i \neq j$ and $q \neq i, j$ $NE(n_i)$ = network efficiency of enterprise resource i p_{iq} = proportion of ith enterprise resource's tie strength to qth enterprise resource's tie strength (interaction with qth enterprise resource divided by the sum of ith contacts) m_{jq} = marginal strength of jth enterprise resource's contact in relation with qth enterprise resource's contact (interaction with qth contact divided by the strongest of jth relationship with anyone)	$NE(n_i) = \sum_j \left[1 - \sum_q p_{iq}, m_{jq}\right]$

of employees, and after purifying via threshold values for each parameter, we insert them into knowledge sharing matrices function and interest.

$$F = \begin{bmatrix} 1 & 1 & 0 & 0 & 0 & 0 \\ 1 & 1 & 0 & 0 & 0 & 0 \\ 0 & 0 & 1 & 1 & 1 & 0 \\ 0 & 0 & 1 & 1 & 1 & 1 \\ 0 & 0 & 1 & 1 & 1 & 0 \\ 0 & 0 & 0 & 1 & 0 & 1 \end{bmatrix}$$

$$I = \begin{bmatrix} 1 & 0 & 1 & 1 & 1 & 0 \\ 0 & 1 & 1 & 1 & 0 & 0 \\ 1 & 1 & 1 & 0 & 0 & 1 \\ 1 & 1 & 0 & 1 & 1 & 1 \\ 1 & 0 & 0 & 1 & 1 & 1 \\ 0 & 0 & 1 & 1 & 1 & 1 \end{bmatrix}$$

Then, using the mathematical relations, we cluster the employees to enterprise resources.

$$F \times I = \begin{bmatrix} 1 & 0 & 0 & 0 & 0 & 0 \\ 0 & 1 & 0 & 0 & 0 & 0 \\ 0 & 0 & 1 & 0 & 0 & 0 \\ 0 & 0 & 0 & 1 & 1 & 1 \\ 0 & 0 & 0 & 1 & 1 & 0 \\ 0 & 0 & 0 & 1 & 0 & 1 \end{bmatrix}$$

The enterprise resources initiate the KS process. Meanwhile, the KDP assessment program evaluates, integrates, deposits and guides the enterprise resources achievements during the KS process. These activities are done by an expert observer who is knowledgeable about the research methodology. Enterprise resources need to interact, to exchange knowledge and information due to the variety of the enterprise resources' expertise. These interactions are recorded by the supervisor of each enterprise resource. Now, the CNIAM consisting of three elements of network range, tie strength and network efficiency is worked out. We studied the enterprise resources for three months. The data from the interactions were recorded by the supervisors to be used for CNIAM computations as given in Table 8.2.

These network measures help to find the more active enterprise resource and the enterprise resource with more information exchanges for incentive purposes. The results showed that enterprise resource 1 was more active than other enterprise resources. To evaluate the enterprise resources' KS, experts analyzed the solutions and scored them according to their standard measures. To establish a measure of KS, the three items were added into a scale from 1 (very low) to 21 (very high). Note those enterprise resources one, two and three presented eight, four and five sharing, respectively. This way, the KS measure of each enterprise resource is evaluated, and the practical solutions are chosen to be employed by the company.

TABLE 8.2
The Results of the CNIAM Computations

Enterprise Resource *i*	$NR(n_i)$	$TS(n_i)$	$NE(n_i)$
1	40%	47	0.51
2	27%	22	0.15
3	33%	31	0.34

CONCLUDING REMARKS

We investigated different structural aspects of enterprise resources' network organization and their knowledge sharing within a KDP. Initially, a pilot group of people in an organization was selected. This group was evaluated through KS parameters using a questionnaire. Considering the questionnaire data, a decision maker configured the KS matrix by a scoring technique. The clustering was performed applying the KS matrix. The pilot group was divided into some research enterprise resources. The tasks were submitted to the enterprise resources and their progress in handling the KS was assessed through a CNIAM. The advantages of such programs are continuous monitoring, gradual knowledge sharing in an organization, the involvement of an organization's employees in the sharing process, updating employees' KS attributes, and a comprehensive network interaction assessment method to guarantee continuous improvement. The applicability of the proposed KS matrix model was worked out in a case study.

REFERENCES

1. Jackson SE, Chuang C-H, Harden EE, Jiang Y, Joseph JM. Toward developing human resource management systems for knowledge-intensive teamwork. In: J. M. Joseph (Ed.), *Research in personnel and human resources management* (vol. 25, pp. 27–70). Amsterdam: JAI; 2006.
2. Arthur JB, Huntley CL. Ramping up the organizational learning curve: Assessing the impact of deliberate learning on organizational performance under gainsharing. *Acad Manag J.* 2005;48(6):1159–70.
3. Collins CJ, Smith KG. Knowledge exchange and combination: The role of human resource practices in the performance of high-technology firms. *Acad Manag J.* 2006;49(3):544–60.
4. Mesmer-Magnus JR, DeChurch LA. Information sharing and team performance: A meta-analysis. *J Appl Psych.* 2009;94(2):535–46.
5. Mayer RC, Gavin MB. Trust in management and performance: Who minds the shop while the employees watch the boss? *Acad Manag J.* 2005;48(5):874–88.
6. Parise S. Knowledge management and human resource development: An application in social network analysis methods. *Adv Dev Hum Res.* 2007;9(3):359–83.
7. Kwok SH., Gao S. Attitude towards knowledge sharing behavior. *J Comp Inf Sys.* 2005;46(2):45–51.
8. Perry-Smith JE. Social network ties beyond nonredundancy: An experimental investigation of the effect of knowledge content and tie strength on creativity. *J Appl Psych.* 2014;99(5):831–46.
9. Shang Y. An agent based model for opinion dynamics with random confidence threshold. *Comm Nonl Sci Num Sim.* 2014;19(10):3766–77.
10. Leonardi PM. Social media, knowledge sharing, and innovation: Toward a theory of communication visibility. *Inf Sys Res.* 2013;25(4):798–816.
11. Fazlollahtabar H, Shirazi B, Porramezan Ganji A. A framework for knowledge sharing of enterprise resources. *Int J Inf Comput Sci (IJICS)*, 2015;4:9–18.
12. Mahdavi I, Fazlollahtabar H, Mahdavi-Amiri N, Arabmaghsudi M, Yahyanejad MH. A virtual intelligent creativity matrix for Employees Clustered Interactivity Network with knowledge development program. *Int J Knowl-Based Organ (IJKBO).* 2014;4(1):65–79.

9 Knowledge Sharing and Organizational Culture

INTRODUCTION

Knowledge management (KM) is not a new phenomenon; human civilizations employed knowledge retention and transfer from one generation to the next to perceive the past and forecast the future. Thirst for knowledge, in the modern dynamic and complex commercial environments, caused firms to spread in amplitude and profundity. Knowledge is changing and disseminating out of the organization. Information technology (IT) and the internet also generate new challenges in knowledge creation, preservation and sharing in KM. Meanwhile, subject to organizational structure and culture, in line with sharing this dynamic knowledge among modern organizations can be very determinant and effective. It is likely that lack of attention to these factors has been the cause of failure in the learning projects of organizations and even a barrier against the implementation of this project, and consequently no knowledge sharing has been done in the organization, which has deprived it of very dynamic competitive advantage. Nowadays, knowledge sharing in KM following learning in the organization is one of the essential subjects in organizations, because in today's world that is fast and on the rise, only dynamic organizations can cope with increasing changes and achieve prosperity and growth.

Several models have been developed for relating knowledge and organizational affairs. The aim of the autopoietic model of knowledge was to act as a common foundation for KM to overcome the numerous KM failures, highlighted by the literature, attributed to inaccurate or constantly debated definitions of knowledge. Parboteeah and Jackson [1] seek to evaluate an autopoietic model.

Organizations need to measure the degree to which they are endowed with the capability to manage customer knowledge effectively to foster innovation. Belkahia and Triki [2] aimed to propose a measurement scale of the customer knowledge enabled innovation (CKEI) capability. The CKEI scale was considered as a barometer allowing organizations to evaluate to what extent they are endowed with the capacity of co-creating value with their customers. The proposed CKEI scale was believed to provide managers with the opportunity to monitor their innovative capability regularly and be close to their customers.

Belso-Martínez et al. [3] aimed to address a central question in strategy: how do firms' internal resources mediate the effect of the external resources on the firms' performance? First, the study used only two well-known internal resources and capabilities indicators. Second, the study applied a strict and simple measure to the growth of innovative firms. Third, another limitation of the research related to the sample and population of companies. The study showed that the partial mediating effect exercised by internal resources and capabilities on growth becomes more intense when new firms benefit from cluster location. This study represented a new step toward closing the analytical gap in the existing literature on the potential interactions between external resources and a new firm's internal attributes and their combined effects on performance.

Based on the definition and characteristic analysis, Zhou and Chen [4] explored the proposal of a formation mechanism of knowledge rigidity, which is constituted by the effects of three precipitating factors: time effectiveness of knowledge, reinforcing effectiveness and sunk cost effect in the knowledge selection mechanism. This chapter provides theoretical support to realize knowledge rigidity in KM practice. Three indicators were proposed to evaluate the rigidity and suggestions for action were given to help control knowledge rigidity in firms.

LITERATURE REVIEW

The work of organizations with KM should focus on attracting knowledge repositories and knowledge transfer without an implicit or internal structure to explicit or external structured knowledge and be considered transforming individual knowledge into organizational knowledge. This issue can not only be explained by the need of organizations for the better management of knowledge with the establishment of core competencies for individuals and performance indexes through the identification of intangible assets, but also organizations try to become an innovative and learning organization with a knowledge-sharing culture. Despite the importance of knowledge as a most strategic resource, there is little empirical research in this area about how that knowledge is created within the organization and what background helps in this process. There is convincing evidence that theories and knowledge-creation measures can be used in different organizations. According to the dynamic theory of organizational knowledge creation, knowledge occurs in dynamic social interaction. Thus, organizational knowledge creation meant to create an appropriate context for this process. Based on the theory of knowledge-based firms, organizations are more effective in the markets in creating and managing the knowledge. Despite this fact, the issue has received little attention in industry, universities and other organizations. They are still at the stage of acquisition and storage of information, and they are designed for the same purpose, while the evidence shows that there is not a linear relationship between the acquisition of information and knowledge creation and acquisition, and storage of information is only the first step [5]. According to the understanding of importance, the relationships as a main factor in the process of knowledge creation and knowledge sharing, organizational capabilities that lead to the more interaction between members are examined and that these elements are studied from the dimension of the organizational structure and culture. It must be understood that based on this knowledge sharing, a dynamic and learning organization is formed. The following factors can be identified from the dimension of the structure that can be effective in the process of knowledge sharing among teams with the same goals:

The concentration of power (implicit in organization structure), communications and conflicts within the organization, or organizational complexity (implicit in organization structure), formality and type of leadership (implicit in organization structure).

The second dimension is the organizational culture that is considered as one of the most important factors in the organization because it deals directly with individuals and staff, and consequently with their values, beliefs and attitudes. Organizational culture is a contextual variable that affects all members of the organization at different levels and therefore appropriate understanding of this structure is important for managing the organization and producing effective work. Cheung has cited the following factors as effective factors in the organizational culture in research that he has conducted about finding an organizational culture framework in the case of construction companies in Hong Kong:

- Preparation and goal attainment;
- Team path;
- Coordination and integration;
- Emphasis on performance;
- Innovation trend;
- Member's partnership;
- The orientation of compensation or reward [6].

A knowledge-sharing culture is one of the most important factors that should be considered because knowledge sharing occurs if the organizational culture supports it. Staff share their ideas and insights with others in the organization that has a culture of knowledge sharing. Wang et al. [7] expressed in a study of knowledge-sharing subjects that the success of KM projects depends on knowledge sharing. Based on the literature review, there is a framework for understanding the research and knowledge-sharing development that includes five areas: organizational concepts, characteristics between individuals and teams, cultural characteristics, individual characteristics and motivational factors. This research shows that organizations should have paid more attention to this issue and that cultural characteristics are effective in the development of human resources characteristics, which is one of the effective factors in knowledge sharing. To achieve this goal, a general feature cannot be used to facilitate sharing between the various communities, but organizations need to regulate and offer a type of appropriate encourager for their cultural areas [7]. Knowledge sharing involves sharing information, ideas, suggestions and specialties between individuals in an organization and refers to the sharing rate of knowledge resources within task and performance boundaries and that this same capability can change business processes substantially. Information sharing not only facilitates inter-task interactions but also the sharing of knowledge repositories among participants in organizational processes and this same issue leads to collaboration and comprehensive understanding of a process.

The distribution and transfer of knowledge has an alternative descriptive role for business processes that transfer and disseminate knowledge among the members of an organization or colleague groups [8]. At this stage, they should be careful that distributed knowledge be offered as appropriate, useful, interpretive and comprehensible. Knowledge distribution channels can be formal or informal so that informal channels can accelerate the process of knowledge socialization and these types of channels are very suitable for small organizations. While knowledge distribution through formal channels, such as training, supports wider distribution of knowledge and is more suitable for the concept-based and specialty-based knowledge in large organizations. Pimchangthong et al. [9] dealt with examining effective factors on KM in industrial plants in Thailand. In this study, the factors studied included technology infrastructure, human resources, organizational culture and knowledge sharing. Data are distributed and collected through questionnaires among 400 employees of ten industrial plants in Thailand and analyzed using multivariate linear regression and multivariate correlation. The questionnaire was designed as two sections; the first section consisted of 20 questions about factors affecting KM with the titles of technology infrastructure (X1), human resources (X2), knowledge sharing (X3)) and organization culture (X4); and the second section consisted of 20 questions about KM processes with the titles of discovery (Y1), absorption (Y2), sharing (Y3) and application (Y4). They obtained the following regression equation about the correlation and regression calculations.

$$\text{Discovery:}\quad Y1 = 1.265 + 0.136X1 + 0.155X2 + 0.106X3 + 0.192X4$$

$$\text{Capture:}\quad Y2 = 1.322 + 0.149X1 + 0.113X2 + 0.111X3 + 0.190X4$$

$$\text{Sharing:}\quad Y3 = 1.907 + 0.219X1 + 0.112X2 + 0.654X4$$

$$\text{Application:}\quad Y4 = 1.636 + 0.120X1 + 0.143X2 + 0.221X4$$

The findings also indicate that knowledge sharing does not have any influence on the two KM processes (sharing and application). It means that there is still resistance to sharing knowledge among staff. As a result, to maintain the organization's tacit knowledge and develop explicit knowledge, the organization must find ways to motivate staff to share knowledge. In this study, the most effective factor in knowledge management is organizational culture. Lindner and Wald [10] dealt with the examination of KM success factors and consequently knowledge sharing in temporary organizations. The depth of the growing knowledge, work content and business project increases the need for KM in temporary organizations and between them. Although with this research several potential success factors were identified for KM projects, focus on a case or a variety of projects was limited by the ability to generalize the results, the relative importance of various factors and insufficient analysis of their common relationship. Therefore, in this study, the main factors affecting the effectiveness of KM projects were examined simultaneously in a large sample of different industries and types of projects. Here, it cannot be confirmed whether some factors affecting the project KM derived from previous studies in different research settings are effective or not. Some of these factors can be found in the general research about KM in permanent organizations. Although some factors have particular importance in temporary organizations, they are employed to compensate for the lack of uniformity and for bridging the gaps between different projects, regarding time, place, distribution of tasks and project individuals deliberately. The influences of cultural factors on the success of project knowledge management especially, are in line with the findings of previous research [11]. An understanding of KM can be expanded in the project environment by showing that knowledge culture is the most important success factor. Although not only the softer factors such as the culture and commitment of top management are necessary for successful transfer of knowledge between and among the various organizations, they must be supplemented with ICT systems that support communications and effective storage and retrieval of knowledge in a temporary project environment. It is not mere access that is essential but also the quality and usefulness. Here multi-project project management organizations and especially the role and organization of project management offices are expressed as the third main success factor. In addition, several other factors like the processes of project knowledge and project knowledge organization have a positive impact on the effectiveness of project knowledge management. Overall, this is the influence of several factors that lead to a successful transfer of knowledge within and between projects from temporary organizations to permanent organizations. The above study based on broad experience shows that the effectiveness of managing knowledge relies on a set of project management practices. In brief, research is designed based on three main dimensions:

1. Organization and processes (structure);
2. Systems and ICT;
3. Culture and leadership of theories.

It proceeds to collect and investigate the necessary information by distributing questionnaires among 496 people in various roles: project managers, leaders, staff, and workers that were members of a German association for project management. The selection of projects tried to sample different activities consisting of productive, software, transport, financial services and the questionnaire used was of a Likert-type scale. We offer more about the relationship between the dependent variable (effect of project KM) and its independent variables and also the result of the studied model that is calculated by the PLS 2.0 software. Finally, the author reaches a conclusion by examining tables and confirming or rejecting the available hypothesis that a factor of culture (as one of the most important factors), organization and ICT infrastructures are effective in achieving the goals of KM in temporary organizations.

Naftanaila [12] dealt with the examination of KM success factors in the project environment. The factors analyzed in this study consist of confidence among staff, team members, project culture, individuals' values and beliefs and the motives of the individuals involved in the project. The researcher studied their impact on the transfer and sharing of studied knowledge by examining each of these components and at the end of this study stated that extant conditions in the projects and culture, norms and values of projects can be a barrier to accelerate the issue of knowledge sharing and transfer. Meanwhile, the inertia force of an organization and its individuals can be among the barriers to knowledge sharing and transfer.

Yap et al. [13] dealt in another study with effective factors on KM methods in extra multimedia organizations. Here researchers dealt with the examination of four main factors affecting KM: culture, IT, organization structure and staff among several extra multimedia companies in Malaysia. Then, using variance analysis (ANOVA) and data mean, studied the relationship between these factors and age, years of attendance in the organization, experience and staff job design. They concluded that the demographic factors associated with the knowledge management factors. Also, without proper attention to structure, organization culture and staff knowledge sharing in the organization cannot be implemented.

Wang et al. [7] expressed about knowledge sharing in another study with this title: *Knowledge Sharing: A Review and Directions for Future Research*, that the success of KM projects depends on sharing knowledge. This study examines qualitative and quantitative studies of knowledge sharing at the individual level. Based on the literature study there is a framework for the understanding of research and development of knowledge sharing that includes five areas: organizational concepts, characteristics between individuals and teams, cultural characteristics, individual characteristics and motivational factors. Results of this research can be stated as follows: culture, with emphasis on reliability and innovation, has a direct impact on management knowledge sharing and also the impact of behavior is indirectly effective on the attitudes of managers. Research shows that management and support of supervisors play a very significant role in the success of KM and knowledge sharing [14]. Managers need to have the necessary support for providing a bonus and encourage staff to share knowledge in organizations. The study illustrates the importance of increasing self-confidence among individuals in knowledge sharing. This research shows that organizations should have paid more attention to the issue that cultural characteristics are effective in the development of human resources characteristics and that it is one of the effective factors in knowledge sharing. To achieve this goal, a general feature cannot be used to facilitate sharing between the various communities but, organizations need to regulate and offer a type of appropriate encourager for their cultural areas. Chang [15] dealt with the examination of vital success factors in knowledge management and the project of framework classification in the Taiwanese government, and therefore uses research based on the examination of effective factors in the success of KM projects, and consequently knowledge sharing, that is an inseparable part of the subject. Hence, the 181 questionnaires distributed among the participants consist of effective factors in the success of KM and some of them are listed below:

- Organizational goals;
- Main competence;
- Human resources;
- Process management;
- Explicit knowledge;
- IT architecture;
- Knowledge activities;
- Optimization;
- Open-ended questions;
- Demographic information.

Then, based on the responses received and their examination using statistical methods it was concluded that effective factors are divided into two sections:

- Dimensions of the KM process – divided into organizational values and mission, IT application, KM documentation, structure and process management and human resources assets;

- Dimensions of KM performance – divided into evolution and acquisition of knowledge, business performance, knowledge sharing and value added.

After the expression of theoretical subjects in KM with research regarding 12 large French companies, Kimble and Bourdon [16] dealt with the characteristics of similar work teams and their organization's structure and texture by means of an interview with their chief knowledge officer. In this subject, they refer to issues such as an individual's behavior in similar work teams and their resistance to sharing knowledge. A method of interaction and information exchange (information repositories and information networks between staff) is expressed. The creation of organizational culture is named based on more sociability of individuals with each other to help create task, interests and common goals for more sharing of knowledge among staff, and in its results names systems to help KM, with the support of the chief officer as one of the most important tools for knowledge sharing, and finally, KM upgrade. Generally, in this research, the subject of culture and individuals' psychological issues in a similar work team, as well as the organization's structure and texture, are expressed as effective factors in creating and upgrading an organization's KM. The results emphasize some "success factors" of social knowledge management. Two types of factors in particular, are introduced to encourage knowledge sharing:

1. Characteristics of a similar work team;
2. Organizational content.

The aim of this study was the insight that consumers believed the factors that have the most influence on the success of a similar work team are based on KM approaches. No wonder, the survey confirmed the importance of a similar work team for KM and showed that human factors are an essential component in the development of systems to help knowledge management. Another study has been conducted in descriptive, analytical and cross-sectional methods with the target of study the relationship between structural and cultural factors of organization and knowledge management strategy in the public health training centers of Tehran University of Medical Sciences in Iran. The study population comprised the staff of the public health training centers of Tehran University of Medical Sciences (nine centers) and up 200 of them were chosen by random sampling; the data gathering tool was a questionnaire. The results indicate that the status of knowledge management is medium because of formality and high centralization in the organizational structure of the centers researched. Also, the correlation coefficient suggests that there is a significant relationship between the structural dimensions and cultural dimensions of the organization and KM ($p < 0.005$), so that this relationship to formality and centralization is negative (respectively with correlation coefficients of −0.144 and −0.272) and is positive to communications flow, knowledge sharing and continuous learning (respectively with correlation coefficients of 0.162, 0.217 and 0.223).

According to the correlation between structural and cultural factors of an organization with the knowledge management, and the role of knowledge in the quality of the provided medical service, it is recommended to pay necessary and multilateral attention to structural and cultural dimensions of organizations in the establishment of knowledge management at medical centers and prepare the atmosphere for multilateral communication and continuous information exchange by giving as much freedom to the staff as possible.

RESEARCH IMPORTANCE AND NECESSITY

Knowledge is a significant strategic resource for companies in the twenty-first century. Researchers and specialists attempt to find how knowledge resources can be collected and managed effectively to be used as a competitive advantage. So, before embarking on the implementation of KM projects, organizations need to assess organizational subsystems and the available resources to identify the best and the most important strategy. The main bases for the creation and implementation of KM are organizational structure and culture (as one of the most important factors) to achieve the main KM objectives like creation, identification, storage and sharing of knowledge among organization staff. It would help the employees in their learning to become a dynamic organization based on knowledge sharing. This research deals with the study of the effective organization structure and culture factors on knowledge sharing among staff. As employees are the unavoidable aspect of knowledge sharing and also employees are within the organizational structure of a company, thus analyzing the tradeoff would be important. In knowledge sharing, it is important to provide an environment stimulating the employees' tendencies to transfer their knowledge to others. Therefore, strengthening cultural views in the organization is crucial.

RESEARCH HYPOTHESES

According to the motivations given above, this research deals with the influences of organizational structure and culture dimensions on knowledge and information sharing in a company for implementation of KM. The following hypotheses are made to survey the influences:

1. The centralization in the organization *has a significant relationship* with the sharing of knowledge in the organization.
2. The complexity in the organization *has a significant relationship* with the sharing of knowledge in the organization.
3. The formality in the organization *has a significant relationship* with the sharing of knowledge in the organization.
4. The cultural factors in the organization *have a significant relationship* with the sharing of knowledge in the organization.

The following chart presented in Figure 9.1 outlines the conceptual model of the study:

As shown in Figure 9.1, the organizational structure and culture are considered as independent variables, and factors of centralization, complexity, formality, teamwork execution, type of performance appraisal, innovation trend, communication of members, bonus system and confidence among staff are dependent variables, and

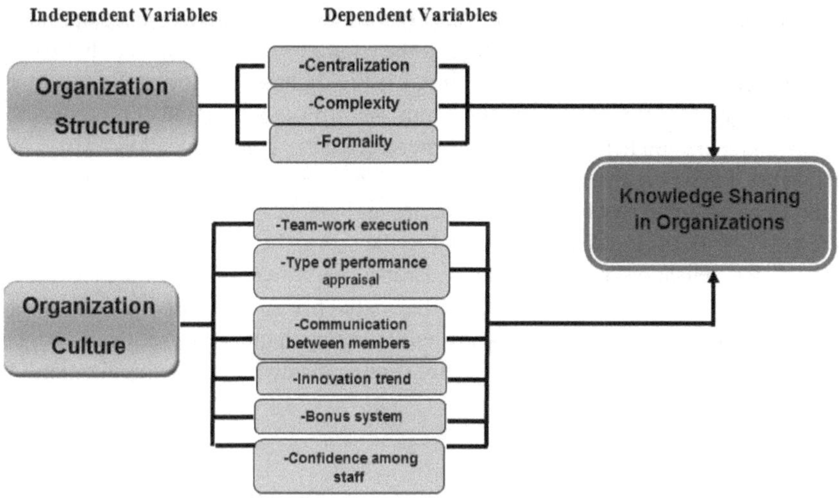

FIGURE 9.1 A conceptual model.

knowledge sharing is also another dependent variable. The aim is to test statistically to find out the effectiveness of the independent variables on the dependent ones. According to Figure 9.1, two levels of independent variables exist, namely, organizational culture and organizational structure. The research method and statistical test employed here are reported in the next section.

RESEARCH METHOD

In this study, a questionnaire is proposed and disseminated via sampling among some managers of all companies and all responses received are analyzed. The study population is selected among managers of food product companies in the industrial park of Lahijan, northern Iran. The samples are analyzed from all the questionnaires (47) and the same number of received responses.

The study questionnaire is the result of an interview with some managers and master administrations about composition, general scheme of the questionnaire and how to design research questions and hypotheses. By using a Likert-type test, it is tried to provide and collect managers' views of some food manufacturing companies in Lahijan industrial park regarding factors affecting organizational structure and culture on the knowledge sharing in the organization. In order to ensure the research tools and verification, the questionnaire is distributed to the pilot and pre-test groups which consist of 30 managers. After examining the validity and reliability according to their opinions, it fixed bugs and the questionnaire was finalized as follows: rate of Cronbach's alpha is $(0.884) > 0.7$, therefore, the questionnaire provides an acceptable measure of reliability. Due to the specialized nature of the questionnaires, they were given to the managers of the study companies to reply to them. The way to interpret this response is that in the section of the organization structure, the higher the score the greater is the complexity, centralization, formality and vice versa. In the section on teamwork, the higher the score the greater is the staff's positive feelings toward the work team, and vice versa. In the section on the type of performance appraisal, the higher the score the more enhanced is the emphasis on performance that means the counseling role and supervisor training is highlighted, the lower the score, the stronger is the traditional performance appraisal. In the section on communication of members, the higher the score the more effective communication exists, and vice versa. In the section on the innovation trend, the higher the score, the greater is the trend for innovation and initiative among members of the organization, and vice versa. About compensation, the higher the score, the greater is the satisfaction through compensation obtained, and vice versa. Also, regarding confidence among staff, the higher the score, the greater is the confidence among individuals, and vice versa; and finally, in the section on knowledge sharing in the organization, the higher the score, the greater is knowledge sharing among members of the organization.

Data Analysis

After collecting the data and entering them into SPSS software, using the software and applying inferential statistics and Spearman correlation calculations, the relationship between the independent variables and the intermediate variables with a research-dependent variable is measured, and by the Friedman test, the priority of the intermediate and independent variables are assessed and prioritized.

The results of research with regard to the demographic data were obtained through descriptive statistics as follows:

> 36.2% of respondents are between the ages of 31 and 35 years, the majority of respondents (74.5%) are men, and 78.7% of respondents are married. Information from managers' resumes is as follows: 10.6% with diploma degree, 10.6% with associate degree, 70.2% with B.A. degree, 8.5% with M.A. degree and/or higher; and with regard to experience, 57.4% of respondents have two to five years' experience.

The research results indicate the following subjects by inferential statistics:

The first hypothesis: Centralization in the organization has a significant relationship with the sharing of knowledge in the organization.

In the test of the first hypothesis, the correlation between the two variables of the complexity in the organization and knowledge sharing in the organization is measured; as a result according to the correlation coefficient −0.815 and sig 0.0 that is lower than value of 0.01, H_0 hypothesis will be rejected, so it can be said that there is a significant relationship between centralization in the organization and knowledge sharing in the organization; therefore, with the increase of centralization in the organization, knowledge sharing in the organization is reduced, and vice versa (Table 9.1).

The second hypothesis: Complexity in the organization has a significant relationship with the sharing of knowledge in the organization.

TABLE 9.1
Spearman Correlation Test of the First Hypothesis

Correlations

			Centralization in Organizations	Knowledge Sharing in Organizations
Spearman's rho	Centralization in organizations	Correlation coefficient	1.000	−.815**
		Sig. (2-tailed)	.	.000
		N	47	47
	Knowledge sharing in organizations	Correlation coefficient	−.815**	1.000
		Sig. (2-tailed)	.000	.
		N	47	47

** Correlation is significant at the 0.01 level (2-tailed).

In the second hypothesis test, the correlation between the two variables of centralization in the organization and knowledge sharing in the organization is measured; as a result according to the correlation coefficient −0.755 and sig 0.0 that is lower than value of 0.01, H_0 hypothesis will be rejected, so it can be said that there is a significant relationship between complexity in the organization and knowledge sharing in the organization; therefore with the increase of complexity in the organization, knowledge sharing in the organization is reduced, and vice versa (Table 9.2).

The third hypothesis: Formality in the organization has a significant relationship with the sharing of knowledge in the organization.

TABLE 9.2
Spearman Correlation Test of the Second Hypothesis

Correlations

			Complexity in Organizations	Knowledge Sharing in Organizations
Spearman's rho	Complexity in organizations	Correlation coefficient	1.000	−.755**
		Sig. (2-tailed)	.	.000
		N	47	47
	Knowledge sharing in organizations	Correlation coefficient	−.755**	1.000
		Sig. (2-tailed)	.000	.
		N	47	47

** Correlation is significant at the 0.01 level (2-tailed).

TABLE 9.3
Spearman Correlation Test of the Third Hypothesis

Correlations

			Formality in Organizations	Knowledge Sharing in Organizations
Spearman's rho	Formality in organizations	Correlation coefficient	1.000	−.859**
		Sig. (2-tailed)	.	.000
		N	47	47
	Knowledge sharing in organizations	Correlation coefficient	−.859**	1.000
		Sig. (2-tailed)	.000	.
		N	47	47

** Correlation is significant at the 0.01 level (2-tailed).

In the test of the third hypothesis, the correlation between the two variables of the formality in the organization and knowledge sharing in the organization is measured; as a result according to the correlation coefficient −0.859 and sig 0.0 that is lower than value of 0.01, H_0 hypothesis will be rejected, so it can be said that there is a significant relationship between formality in the organization and knowledge sharing in the organization; therefore with the increase of formality in the organization, knowledge sharing in the organization is reduced, and vice versa (Table 9.3).

The fourth hypothesis: The cultural factor in the organization has a significant relationship with the sharing of knowledge in the organization.

According to the six cultural factors in this research, engaging in teamwork, type of performance appraisal, communication between members, innovation, compensation and confidence among staff, at first the correlation between these factors and knowledge sharing in the organization is examined, and then in general, influence of cultural factors on knowledge sharing in the organization is discussed.

1. The correlation between the two variables of engaging in teamwork in the organization and knowledge sharing in the organization is measured; as a result according to the correlation coefficient 0.812 and sig 0.0 that is lower than value of 0.01, H_0 hypothesis will be rejected, so it can be said that there is a significant relationship between engaging in teamwork in the organization and knowledge sharing in the organization; therefore with the increase of teamwork in the organization, knowledge sharing in the organization is enhanced, and vice versa (Table 9.4).
2. The correlation between the two variables of the type of performance appraisal in the organization and knowledge sharing in the organization is

TABLE 9.4
Spearman Correlation Test of the Fourth Hypothesis – The First Factor

Correlations

			Teamwork in Organizations	Knowledge Sharing in Organizations
Spearman's rho	Teamwork in organizations	Correlation coefficient	1.000	.812**
		Sig. (2-tailed)	.	.000
		N	47	47
	Knowledge sharing in organizations	Correlation coefficient	.812**	1.000
		Sig. (2-tailed)	.000	.
		N	47	47

**. Correlation is significant at the 0.01 level (2-tailed).

measured; as a result according to the correlation coefficient 0.788 and sig 0.0 that is lower than value of 0.01, H_0 hypothesis will be rejected, so it can be said that there is a significant relationship between type of performance appraisal in the organization and knowledge sharing in the organization; therefore leading to more advanced models of performance appraisal in the organization that is based on consulting role and supervisor training, knowledge sharing in the organization is enhanced, and with the traditional performance appraisal in the organization that focuses on the output of the appraisal, knowledge sharing in the organizations is reduced (Table 9.5).

TABLE 9.5
Spearman Correlation Test of the Fourth Hypothesis – The Second Factor

Correlations

			Type of Performance Appraisal in Organizations	Knowledge Sharing in Organizations
Spearman's rho	Type of performance appraisal in organizations	Correlation Coefficient	1.000	.788**
		Sig. (2-tailed)	.	.000
		N	47	47
	Knowledge sharing in organizations	Correlation Coefficient	.788**	1.000
		Sig. (2-tailed)	.000	.
		N	47	47

** Correlation is significant at the 0.01 level (2-tailed).

TABLE 9.6
Spearman Correlation Test of the Fourth Hypothesis – The Third Factor

Correlations

			Knowledge Sharing in Organizations	Knowledge Sharing in Organizations
Spearman's rho	Communication between members in organizations	Correlation coefficient	1.000	.869**
		Sig. (2-tailed)	.	.000
		N	47	47
	Knowledge sharing in organizations	Correlation coefficient	.869**	1.000
		Sig. (2-tailed)	.000	.
		N	47	47

** Correlation is significant at the 0.01 level (2-tailed).

3. The correlation between the two variables of communication between members in the organization and knowledge sharing in the organization is measured; as a result according to the correlation coefficient 0.869 and sig 0.0 that is lower than value of 0.01, H_0 hypothesis will be rejected, so it can be said that there is a significant relationship between communication between members in the organization and knowledge sharing in the organization; therefore with the increase of effective communication between members in the organization, knowledge sharing in the organization is enhanced, and vice versa (Table 9.6).
4. The correlation between the two variables of innovation in the organization and knowledge sharing in the organization is measured; as a result according to the correlation coefficient 0.843 and sig 0.0 that is lower than value of 0.01, H_0 hypothesis will be rejected, so it can be said that there is a significant relationship between innovation in the organization and knowledge sharing in the organization; therefore with the increase of innovation in the organization, knowledge sharing in the organization is enhanced, and vice versa (Table 9.7).
5. The correlation between the two variables of compensation in the organization and knowledge sharing in the organization is measured; as a result according to the correlation coefficient 0.831 and sig 0.0 that is lower than value of 0.01, H_0 hypothesis will be rejected, so it can be said that there is a significant relationship between compensation in the organization and knowledge sharing in the organization; therefore with the increase of compensation in the organization, knowledge sharing in the organization is enhanced, and vice versa (Table 9.8).
6. The correlation between the two variables of confidence among staff in the organization and knowledge sharing in the organization is measured;

TABLE 9.7
Spearman Correlation Test of the Fourth Hypothesis – The Fourth Factor

Correlations

			Innovation Trend in Organizations	Knowledge Sharing in Organizations
Spearman's rho	Innovation trend in organizations	Correlation coefficient	1.000	.843**
		Sig. (2-tailed)	.	.000
		N	47	47
	Knowledge sharing in organizations	Correlation coefficient	.843**	1.000
		Sig. (2-tailed)	.000	.
		N	47	47

** Correlation is significant at the 0.01 level (2-tailed).

TABLE 9.8
Spearman Correlation Test of the Fourth Hypothesis – The Fifth Factor

Correlations

			Bonus System in Organizations	Knowledge Sharing in Organizations
Spearman's rho	Bonus system in organizations	Correlation Coefficient	1.000	.831**
		Sig. (2-tailed)	.	.000
		N	47	47
	Knowledge sharing in organizations	Correlation Coefficient	.831**	1.000
		Sig. (2-tailed)	.000	.
		N	47	47

** Correlation is significant at the 0.01 level (2-tailed).

as a result according to the correlation coefficient 0.751 and sig 0.0 that is lower than value of 0.01, H_0 hypothesis will be rejected, so it can be said that there is a significant relationship between confidence among staff in the organization and knowledge sharing in the organization; therefore with the increase of confidence among staff in the organization, knowledge sharing in the organization is enhanced, and vice versa (Table 9.9).

According to the test by six factors in the organizational culture and with the results it was determined that organizational culture is an important and constructive factor in knowledge sharing in the organization and consequently creates a learning

TABLE 9.9
Spearman Correlation Test of the Fourth Hypothesis - The Sixth Factor

Correlations

			Confidence Among Staff in Organizations	Knowledge Sharing in Organizations
Spearman's rho	Confidence among staff in organizations	Correlation coefficient	1.000	.751**
		Sig. (2-tailed)	.	.000
		N	47	47
	Knowledge sharing in organizations	Correlation coefficient	.751**	1.000
		Sig. (2-tailed)	.000	.
		N	47	47

** Correlation is significant at the 0.01 level (2-tailed).

organization. As the Spearman correlation test above shows, coefficients of the direct impact of cultural factors on knowledge sharing in the organization are relatively high and that indicates that culture is an important factor in knowledge sharing. With the provided results, it can be said that the fourth hypothesis is also accepted and consequently four research hypotheses are confirmed.

Variable Rankings Based on Friedman Test

In this section, the Friedman statistical test is used to investigate the priority of structural and cultural factors for knowledge sharing in the organization. The statistical hypotheses of this test are:

H_0: Ranks of importance for each of the nine effective factors in knowledge sharing are equal.
H_1: There is a significant difference in at least two ranks of importance for the nine effective factors in knowledge sharing.

Statistical analysis results of questionnaire data are given in Table 9.10 by SPSS software regarding the above hypotheses.

According to this test, the significant level of the Friedman statistical analysis is lower than the error range of 0.05. Consequently, it can be said the sample data has not offered a good reason to accept H_0 in significance level of 95%, therefore the H_1 hypothesis is confirmed. In other words, with the confidence of 95% it can be said that there is a significant difference in at least two ranks for the nine effective factors in knowledge sharing. The relative importance rankings of factors are presented in Table 9.11.

TABLE 9.10
Friedman Test

Test Statistics	
N	47
Chi-Square	70.101
Df	8
Asymp. sig.	.000

TABLE 9.11
The Relative Importance Ranking of Factors

Factors	Average Ratings	Important Priority
Teamwork	6.39	First
Centralization	5.73	Second
Innovation trend	5.66	Third
Confidence among staff	5.37	Fourth
Bonus system	5.16	Fifth
Communication of members	5.06	Sixth
Formality	4.93	Seventh
Complexity	4.43	Eighth
Type of performance appraisal	2.27	Ninth

As is indicated in the above table, structural factors of centralization in the organization, formality in the organization, complexity in the organization, respectively, have the most converse influence on knowledge sharing in organizations. Also, the cultural factors of teamwork, innovation, confidence among staff, compensation, communication of members and type of performance appraisal, respectively, have been allocated the most direct impact on knowledge sharing.

CONCLUDING REMARKS

According to verification of the first hypothesis, it is determined that centralization in the organization, that is one of the organizational structure dimensions, has a significant and converse relationship with knowledge sharing in the organization leading to centralized and ordered decisions in the high levels of the organization, lack of dynamism, and consequently apathy and closing of communication channels for knowledge sharing. According to verification of the second hypothesis, it is determined that complexity in the organization, that is one of the organizational structure dimensions, has significant and converse relationship with knowledge sharing in the organization and is effective. It is due to the existence of different levels of hierarchy, positions and titles, and consequently making the organization more complex, and creates a type of competition among staff leading to resistance in knowledge sharing among the organization's staff. According to verification of the third hypothesis, it is determined that formality in the organization, that is one of the organizational structure dimensions, has significant and converse relationship with knowledge sharing in the organization. Due to the restrictive legislation, ground rules and policies in the organization, information and knowledge sharing among staff is not allowed or at least is very limited, and opportunities for knowledge sharing among staff will be very few. According to verification of the fourth hypothesis, it is determined that cultural factors have a significant and direct relationship with knowledge sharing in the organization and are effective. Now, about the cultural variables discussed in this study which were tested, we examined the effects of these variables. The first cultural characteristic discussed is teamwork in the organization that has a significant and direct relationship with knowledge sharing in the organization. This can be due to the recognition of colleagues and an atmosphere of collaboration with others. The second cultural characteristic discussed is the types of performance appraisal in the organization that have a significant and direct relationship with knowledge sharing in the organization. In an advanced performance appraisal, there is great emphasis on the counseling role and supervisor training, so supervisors help subordinates to do their jobs better and staff are trained, and this is a kind of inherent knowledge sharing. But, in the traditional performance appraisal methods, supervisors appear in the role of a documentary director about the performance of subordinates for management and control purposes, therefore they are no longer considered as a knowledge-sharing leader. The third cultural characteristic discussed is communication among staff in the organization that has a significant and direct relationship with knowledge sharing in the organization. The more communication among staff in the organization and positive relationships of cooperation among them, the better and easier is the knowledge sharing in the organization, and so they will show less resistance to knowledge sharing among themselves, and vice versa. The fourth cultural characteristic discussed is innovation trends among staff in the organization that have a significant and direct relationship with knowledge sharing in the organization. In organizations where staff have more willingness to innovate equally and are interested in the supply and use of technology, and consequently, to achieve this important object, they try to align their views with others through explaining and sharing their knowledge. The fifth cultural characteristic discussed is compensation in the organization that has a significant and direct relationship with knowledge sharing in the organization. It can be interpreted by staff that the compensation system considers them as a part of the organization and tries to use all their abilities and knowledge to improve knowledge of the other staff and promote the organization. It helps to increase knowledge multiplexing and sharing in organizations. The sixth cultural characteristic discussed is confidence among staff in the organization that has a significant and direct relationship with knowledge sharing in the organization. It is due to the spirit of confidence and trust to create the sharing of knowledge inherently. Individuals trust each other, which leads to a constructive relationship, and consequently to knowledge multiplexing and sharing. As future research, analysis can be directed to other aspects of knowledge management on organizational culture and structure.

REFERENCES

1. Parboteeah P, Jackson TW. Expert evaluation study of an autopoietic model of knowledge. *J Knowl Manag.* 2011;15(4):688–99.
2. Belkahla W, Triki A. Customer knowledge enabled innovation capability: Proposing a measurement scale. *J Knowl Manag.* 2011;15(4):648–74.
3. Belso-Martínez J.A, Molina-Morales FX, Mas-Verdu F. Clustering and internal resources: Moderation and mediation effects. *J Knowl Manag.* 2011;15(5):738–58.
4. Zhou Z, Chen Z. Formation mechanism of knowledge rigidity in firms. *J Knowl Manag.* 2011;15(5):820–35.
5. Adlie F. Model of knowledge creation and sharing in organizations. In: *First National Conference on Knowledge Management*, Iran, February 2008.
6. Cheung S. Towards an organizational culture framework in construction. *I J Proj Manag.* 2011;29(1):33–44.
7. Wang S., Noe R. Knowledge sharing: A review and directions for future research. *Hum Resour Manag Rev.* 2010;20(2):115–31.
8. Akroush MN, Abu-ElSamen AA, Jaradat NA. The influence of mall shopping environment and motives on shoppers' response: A conceptual model and empirical evidence. *Int J Oper Manag.* 2011;10(2):168–98.
9. Pimchangthong D, Tinprapa S. Factors influencing knowledge management process model: A case study of manufacturing industry in Thailand. *World Acad Sci Eng Tech.* 2012;64:588–91.
10. Lindner F, Wald A. Success factors of knowledge management in temporary organizations. *Int J Proj Manag.* 2011;29(7):877–88.
11. Eraqi MI. Strategic marketing as a managerial approach for enhancing the competitiveness of the tourism business sector in Egypt. *Int J Serv Oper Manag.* 2010;7(4):483–500.
12. Naftanaila I. Factors affecting knowledge transfer in project environments. *Rev Int Compar Manag.* 2010;11:834–40.
13. Yap LS, Tasmin R, Rusuli SC, Norazlin H. Factors influencing knowledge management practices among multimedia super corridor (MSC) organizations. *Comm IBIMA.* 2010;2010:1–12.
14. Sony M, Mekoth N. A typology for frontline employee adaptability to gain insights in service customisation: A viewpoint. *I J Serv Oper Manag.* 2012;12(4):490–508.
15. Chang M-Y. The research on the critical success factors of knowledge management and classification framework project in the executive Yuan of Taiwan Government. *Exp Sys Appl.* 2009;36(3):5376–86.
16. Kimble C, Bourdon I. Some success factors for the communal management of knowledge. *I J Inf Manag.* 2008;28(6):461–7.

10 Knowledge Sharing and Learning Capability

INTRODUCTION

Organizational learning is a process through which the organization will learn more knowledge. Such learning means any changes in organizational models which may lead to recovery or maintenance of the organizational function [1]. Jerez-Gomez et al. [2] have also defined organizational learning as a creation, procurement, knowledge transfer and integration capability and modification of organizational behavior to reflect a new position with the purpose of the improvement in organizational function. Templeton et al. [3] believed that organizational learning is a collection of organizational functions such as learning knowledge, distribution and interpretation of information and memory consciously and/or non-consciously with positive effects on organizational change. Learning capability is an important factor for the further growth and innovation of an organization. Organizational learning capability is a collection of resources and/or tangible and intangible skills for which it is also necessary to use competitive advantages. An organizational learning capability is a sign of creation capacity and the combination of ideas in an effective way in contact with various organizational borders and through special managerial methods and innovations [4]. There are different studies for measuring organizational learning capabilities at industrial and non-industrial places [5]. Aghdasi and Khakbaz Bafruei [6] studied the organizational learning levels at different hospitals. Knowledge transfer and integration capability had the highest mean, and after them the systems perspective, openness and experimentation and managerial obligation capabilities. There is another study made by Bhatnagar in India for measuring the organizational learning capability of managers. According to the results, IT managers and multinational companies had the highest rate of organizational learning capability, and engineering managers had the lowest [7].

By reviewing the literature on organizational learning capability, learning capability includes the following ten dimensions: risk-taking, interaction with the external environment, dialogue, participative decision making, managerial commitment, systems perspective, openness and experimentation, knowledge transfer and integration, teamwork, demonstration of mission and goals. Organizational learning capability (OLC) is defined as the organizational and managerial characteristics or factors that facilitate the organizational learning process or allow an organization to learn. And organizational learning is seen as a dynamic process based on knowledge, which implies moving among the different levels of action, going from the individual to the group level, and then to the organizational level and back again. This process stems from the knowledge acquisition by the individuals and progresses with the exchange and integration of this knowledge until a corpus of collective knowledge is created, embedded in the organizational processes and culture. This collective knowledge, which is stored in the so-called organizational memory, has an impact on the type of knowledge acquired and the way in which it is interpreted and shared. What an individual learns in an organization greatly depends on what is already known by the other members of the organization – in other words, on the common knowledge base. In this chapter, we focus on the dimensions of OLC to obtain the most levels of learning capability. Increasing learning capability facilitates the knowledge sharing process in different parts of an organization. We should allow performing the transfer, interpretation and integration of the knowledge in the organization. To obtain the maximum level of learning capability in an organization we should implement each dimension of learning capability in the part that has the greatest effect on learning capability compared with other parts. For this, we cluster dimensions of learning capability in different segments, according to the effect of dimensions in different segments.

ORGANIZATIONAL LEARNING CAPABILITY

The concept of capability is introduced as: "The ability of an organization to learn from its experiences and taking them through times and borders". The learning organization or prescriptive literature mainly focuses on the development of normative models for the creation of a learning organization. Gomez et al. [2] suggested three basic concepts: (1) Knowledge, its acquisition, use, distribution and integration in an organization becomes one of the critical strategic resources, and the base of learning in the organization. Acquiring and distributing knowledge is due to internal changes that may be the results of both conceptual and behavioral levels. (2) Learning capability is based on the existence of a collective ego that helps us to see the organization as a system in which every member should try and cooperate to reach desirable results. (3) Because this type of learning is mostly based on time and resources, the value and stability of competitive advantage are higher. This learning needs an open atmosphere for ideas and high levels of experience. One way of preparing for an open atmosphere is to devote a room to new ideas, improvement and renovation of individual knowledge. Learning capability is a complex multidimensional construct. Although various studies have identified different dimensions or components, most do so from a theoretical point of view, there being very few that actually design a measurement scale based on the dimensions identified. These four dimensions are: managerial commitment, systems perspective, openness and experimentation and knowledge transfer and integration. Chiva et al. [8] developed an OLC measurement instrument that understands OLC as a multidimensional concept, the dimensions of which are: experimentation, risk-taking, interaction with the external environment, dialogue and participative decision making. On the one hand, these five dimensions are essential enablers of the organizational learning process; on the other hand, they represent the OLC of a particular firm.

Organizational Learning Capability Dimensions

Risk-Taking

Risk-taking is understood as the tolerance of ambiguity, uncertainty and errors. A range of activities was proposed to facilitate organizational learning, amongst which is stressed the design of environments that assume risk-taking and accept mistakes.

Accepting or taking risks involves the possibility of mistakes and failure. If the organization aims to promote short-term stability and performance, then success is recommended, since it tends to encourage maintenance of the status quo. According to a rule the benefits brought about by error or risk tolerance, prompting of attention to problems and the search for solutions, ease problem recognition and interpretation, and a variety of organizational responses. Since the appearance of this work, many authors have underlined the importance of risk-taking and accepting mistakes in order for organizations to learn.

Interaction with the External Environment

We define this dimension as the scope of relationships with the external environment. The external environment of an organization is defined as factors that are beyond the organization's direct control of influence. It consists of industrial agents such as competitors, and the economic, social, monetary and political/legal systems. Environmental characteristics play an important role in learning, and their influence on organizational learning has been studied by a number of researchers. Relations and connections with the environment are very important, since the organization attempts to evolve simultaneously with its changing environment. More turbulent environments generate organizations with greater needs and desires to learn. In recent years researchers have stressed the importance of observing, opening up to and interacting with the environment.

Dialogue

Authors from the social perspective, in particular, highlight the importance of dialogue and communication for organizational learning. Dialogue is defined as a sustained collective inquiry into the processes, assumptions and certainties that make up everyday experience. The vision of organizational learning as a social construction implies the development of a common understanding, starting from a social base and relationships between individuals. The chance to meet people from other areas and groups increases learning. By working in a team, knowledge can be shared and developed amongst its members. Easterby-Smith et al. [9] hold that the recent literature is moving away from a vision of an integrating dialogue in which consensus is sought, towards one that seeks pluralism and even conflict. Individuals or groups with different visions who meet to solve a problem or work together create a dialogue community.

Participative Decision Making

Participative decision making refers to the level of influence employees have in the decision-making process. Organizations implement participative decision making to benefit from the motivational effects of increased employee involvement, job satisfaction and organizational commitment.

Managerial Commitment

Management should recognize the relevance of learning, thus developing a culture that promotes the acquisition, creation and transfer of knowledge as fundamental values. Management should articulate a strategic view of learning, making it a central visible element and a valuable tool with an influence on the obtaining of long-term results. Likewise, management should ensure that the firm's employees understand the importance of learning and become involved in its achievement, considering it an active part of the firm's success. Finally, management should drive the process of change, taking the responsibility of creating an organization that can regenerate itself and face up to new challenges.

Systems Perspective

A systems perspective entails bringing the organization's members together around a common identity. The various individuals, departments and areas of the firm should have a clear view of the organization's objectives and understand how they can help in their development. The organization should be considered as a system that is made up of different parts, each with its own function but acting in a coordinated manner. Viewing the firm as a system implicitly involves recognizing the importance of relationships based on the exchange of information and services and infers the development of shared mental models. Inasmuch as organizational learning implies shared knowledge, perceptions and beliefs, it will be enhanced by the existence of a common language and joint action by all the individuals involved in the process. Thus, the presence of a common language favors knowledge integration – a crucial aspect in the development of organizational learning. In this way, organizational learning goes beyond the employees' individual learning and takes on a collective nature.

Openness and Experimentation

Our unit of analysis is generative or double-loop learning, which requires an atmosphere of openness that welcomes the arrival of new ideas and points of view, both internal and external, allowing individual knowledge to be constantly renewed, widened and improved. To create an atmosphere of openness, there needs to be a previous commitment to cultural and functional diversity, as well as a readiness to accept all types of opinions and experiences and to learn from them, avoiding the egocentric attitude of considering one's own values, beliefs and experiences to be better than the rest. Openness to new ideas, coming from inside the organization or from outside it, favors experimentation, an essential aspect of generative learning, since it implies the search for innovative, flexible solutions to current and future problems, based on the possible use of different methods and procedures. Experimentation requires a culture that promotes creativity, enterprise ability and the readiness to take controlled risks, supporting the idea that one can learn from one's mistakes.

Knowledge Transfer and Integration

This dimension refers to two closely linked processes which occur simultaneously rather than successively: internal transfer and integration of knowledge. The efficacy of these two processes rests on the previous existence of absorptive capacity, implying the lack of the internal barriers that impede the transfer of best practices within the firm. Transfer implies the internal spreading of knowledge acquired at an individual level, mainly through conversations and interaction among individuals. Fluid communication relies mainly on the existence of agile information systems that guarantee the accuracy and availability of the information. With regard to dialogue and debate, work teams and personnel meetings can be ideal forums in which to openly share ideas. The main role of work teams in developing organizational learning is frequently underlined in the literature, with particular emphasis placed on multidiscipline and multifunction teams. Team learning places the group above the individual, allowing the transfer, interpretation and integration of the knowledge acquired individually. This integration leads to the creation of a collective corpus of knowledge rooted in organizational culture, work processes and the remaining elements that form the "organizational memory". Thus, the knowledge can be subsequently recovered and applied to different situations, guaranteeing the firm's constant learning in spite of the natural rotation of its members.

Teamwork

In today's complex world, individuals need to help each other accomplish organizational objectives. Structures and systems in the organization need to encourage teamwork and group problem solving by employees and reduce the dependency on upper management. Teams also need to have the ability to work cross-functionally. By working in teams, knowledge can be shared among organizational members, and there is also a better understanding of other individuals, their needs and how they work in different parts of the organization, encouraging knowledge transfer as well.

Demonstration of Mission and Goals

The organization as a whole, and each unit within it, needs to have an articulated purpose. Employees need to understand this purpose and how the work they have contributed to the attainment of the mission of the organization. Also, the organization needs to promote employee commitment to these goals. Researchers have stated that "building a shared vision", especially of a future desired state, creates tension that leads to learning. Employees understand the gap between the vision and the current state and can better strive to overcome that gap [10].

PROBLEM DEFINITION AND MODELING

Organizational learning capability has many dimensions. Here, we have involved ten dimensions of OLC for analysis. We can choose each set of OLC dimensions. To obtain the maximum level of learning capability and facilitating the knowledge-sharing process in different segments of an organization, we should implement each dimension of OLC in its appropriate place. At first, we should calculate the weight of each dimension in each part, to do this; we use k' measuring items that exist in OLC literature. Finally, the dimensions are clustered in different parts of the organization according to their effects and the cost of implementing the elements and presented formulas. Then we have considered k implementing methods for implementing each dimension in each part of the organization. Implementing methods for each dimension are different in different organizations. As people learn in different ways, there are different styles of organizational learning; therefore the organizations select different implementing methods for implementing OLC dimensions in different parts according to features of the industrial environment, adopted strategies, business culture, technology, available resources and history of the company. Implementing methods are determined by the organization's knowledge management. The organization limits the total budget to a maximum value, which is considered to have increased the learning capability of the organization[10].

Our proposed formulas for clustering are described in four steps:

Step 1: Determining $N_{ijkk'}$.

We should measure the weight of implementation item k of dimension i in part j by measure item k'.

$N_{ijkk'}$ gives a number between 0 and 100 which is determined by the knowledge management team.

Step 2: Calculating $W_{ijkk'}$.

According to the amount of $N_{ijkk'}$, we can determine the numerical value of k' by specifying the weight of implementation item k of dimension i in part j. Also, the organization's knowledge management team determines the value of a, b, c, d, α, β and γ in $W_{ijkk'}$.

$$w_{ijkk'} = \begin{cases} 0 & N_{ijkk'} < a \\ \alpha & a \le N_{ijkk'} < b \\ \beta & b \le N_{ijkk'} < c \\ \gamma & c \le N_{ijkk'} < d \\ 100 & N_{ijkk'} \ge d \end{cases} \tag{10.1}$$

Step 3: Calculating W_{ijk}.

(10.2) shows the calculation of the weight of implementation item k of dimension i in part j.

$$w_{ijk} = \sum_{k'} w_{ijkk'} \tag{10.2}$$

If implementation item k of dimension i exists in part j, $x_{ijk} = 1$ and otherwise x_{ijk} would be 0. For measuring the effect of each dimension on the learning capability of the organization in each part, we use measurement items in the literature of the organizational learning capability in Table 10.1. We have k' measurement items and k implementation items for each dimension. At first, to achieve this goal we should examine how much is the amount of measurement item k'. This value is shown in $N_{ijkk'}$. $N_{ijkk'}$ which is a number between 0 and 100 and is determined by the organization's knowledge management team. According to the amount of $N_{ijkk'}$, the organization would determine the numerical value of measurement item, k', to measure the weight of implementation item k of dimension i in part j. The organization's knowledge management team determines the value of a, b, c, d, α, β and γ in the function $W_{ijkk'}$.

Step 4: Mathematical clustering model.

TABLE 10.1
Items Composing the OLC Scale

Dimension	Measurement Items
Risk-taking	People are encouraged to take risks in this organization.
	People here often venture into unknown territory.
Interaction with the external environment	It is part of the work of all staff to collect, bring back and report information about what is going on outside the company.
	There are systems and procedures for receiving, collating and sharing information from outside the company.
	People are encouraged to interact with the environment: competitors, customers, technological institutes, universities, suppliers, etc.
Dialogue	Employees are encouraged to communicate.
	There is a free and open communication within my work group.
	Managers facilitate communication.
	Cross-functional teamwork is a common practice here.
Participative decision making	Managers in this organization frequently involve employees in important decisions
	Policies are significantly influenced by the view of employees.
	People feel involved in main company decisions.
Teamwork and group problem-solving	The current approach of the organization encourages personnel to solve the problems cooperatively, before discussing them with managers.
	We often cannot form unofficial groups to solve the problems of the organization.
	The majority of problem-solving groups are members of different operating environments.
Demonstration of mission and goals	There are widespread support and acceptance of the organization's mission statement.
	I do not understand how the mission of the organization is to be achieved (r).
	The organization's mission statement identifies values to which all employees must conform.
	We have opportunities for self-assessment concerning goal attainment.
Managerial commitment	The managers frequently involve their staff in important decision-making processes.
	Employee learning is considered more of an expense than an investment.
	The firm's management looks favorably on carrying out changes in any area to adapt to and keep ahead of new environmental situations.
	Employee learning capability is considered a key factor in this firm.
	In this firm, innovative ideas that work are rewarded.
Systems perspective	All employees have generalized knowledge regarding this firm's objectives.
	All parts that make up this firm (departments, sections, work team, and individuals) are well aware of how they contribute to achieving the overall objectives.
	All parts that make up this firm are interconnected, working together in a coordinated fashion.
Openness and experimentation	This firm promotes experimentation and innovation as a way of improving the work processes.

(Continued)

TABLE 10.1 (CONTINUED)
Items Composing the OLC Scale

Dimension	Measurement Items
	This firm follows what other firms in the sector are doing, adopting those practices and techniques it believes to be useful and interesting.
	Experiences and ideas provided by external sources (advisors, customers, training firms, etc.) are considered a useful instrument for this firm's learning.
	Part of this firm's culture is that employees can express their opinions and make suggestions regarding the procedures and methods in place for carrying out tasks.
Knowledge transfer and integration	Errors and failures are always discussed and analyzed in this firm, on all levels.
	Employees have the chance to talk among themselves about new ideas, programs and activities that might be of use to the firm.
	In this firm, teamwork is not the usual way to work.
	The firm has instruments (manuals, databases, files, organizational routines, etc.) that allow what has been learned in past situations to remain valid, although the employees are no longer the same.

Equation (10.3) or objective function maximizes the total effects of dimensions that implements an organization's different parts.

$$\text{Max} \sum_i \sum_j \sum_k w_{ijk} x_{ijk} \tag{10.3}$$

Because of limited capital, each organization should invest in dimensions of organizational learning capability which has the greatest effect on learning capability of the organization to achieve the highest level of learning capability.

$$\sum_i \sum_k c_{ijk} \cdot x_{ijk} \leq A_j \qquad \forall j \tag{10.4}$$

This equation certifies that the total implementation cost for activating dimensions in part j is limited to a maximum value.

$$\sum_i \sum_j \sum_k c_{ijk} x_{ijk} \leq B \tag{10.5}$$

This equation ensures that the total budget of the organization which is considered to have increased the learning capability of the organization is limited to a maximum value.

$$\sum_i \sum_k w_{ijk} x_{ijk} \geq M_j \qquad \forall j \tag{10.6}$$

This equation indicates that the total effect of dimensions in each part is limited to a minimum value. When the amount of a measured item reduces, its effect will be reduced and thus the total effect on the learning capability of the organization will be reduced. Since we have limited the total effect to a minimum value in each part of the organization, more dimensions will be active and so the total cost will be increased.

COMPUTATIONAL RESULTS

We have considered an organization including five segments, three measurement items and three implementation items for each dimension. The total budget of the organization is allocated to increase the learning capability, the total implementation cost for implementing dimensions in part *j* and the total effects in each part are shown in Table 10.2. Constant coefficients, α, β, γ, a, b and c for calculation of w_{ijk} are shown in Table 10.3. These coefficients are determined by the organization's knowledge management team. We choose $N_{ijkk'}$ and C_{ijk} as follows: $10 \leq N_{ijkk'} \leq 87$; $1800\$ \leq C_{ijk} \leq 4000\$$. We solve the proposed model using Lingo software. Implemented dimensions of an organization's different parts are shown in Table 10.4.

The dimensions of the organization to be considered in each segment of the firm are shown in Table 10.4 ($(i, j, k) = 1$). The implementing method of dimensions in each part is shown in Table 10.4. We have considered three implementing methods for each dimension ($k = 1, 2, 3$). Implementing methods for each dimension are different in different organizations.

CONCLUDING REMARKS

This chapter proposed a mathematical clustering model to disseminate organizational learning capability indices within the context of knowledge management

TABLE 10.2
Values of Fixed Variables

	The Total Implementation Cost = US$70,000	
Part j	Maximum Cost of Implementing Dimensions in Part j (US$)	The Minimize Effect in Part j
1	16,000	70
2	14,000	75
3	24,000	70
4	14,000	70
5	18,000	80

TABLE 10.3
Values of Constant Coefficients

α	β	γ	*A*	*B*	*C*	*D*
27	45	67	13	38	53	79

TABLE 10.4
Output Results

	x_{ijk}														
k	1					2					3				
j															
i	1	2	3	4	5	1	2	3	4	5	1	2	3	4	5
1	0	1	1	0	0	0	0	0	0	0	0	0	0	0	0
2	1	0	0	0	0	0	0	1	0	0	0	0	0	1	0
3	0	0	0	0	0	1	1	0	0	0	0	0	1	0	0
4	0	0	0	0	1	0	0	0	0	0	1	0	0	0	0
5	0	0	0	1	0	0	0	0	0	1	0	0	0	0	0
6	1	0	0	0	0	0	1	0	0	0	0	0	0	0	0
7	0	0	0	0	0	0	0	0	1	0	1	0	0	0	0
8	0	0	0	1	0	0	0	0	0	1	0	0	0	0	0
9	1	0	0	0	0	0	1	0	0	0	0	0	0	0	0
10	0	0	0	0	0	0	0	0	1	0	0	0	0	0	1

in firms. The mathematical clustering technique has determined the allocation of indices according to their effect on the learning capability to different parts of the organization to obtain the highest level of learning capability in an organization. We developed an algorithm to imply the steps of the clustering method. Computational results confirmed the effectiveness of the model. The findings showed that the learning capabilities that have more effects are in a cluster related to one specific segment of the firm.

REFERENCES

1. Alegre J, Chiva R. Assessing the impact of organizational learning capability on product innovation performance: An empirical test. *Technovation.* 2008;28(6):315–26.
2. Jerez-Gomez P, Céspedes-Lorente J, Valle-Cabrera R. Organizational learning capability: A proposal of measurement. *J Bus Res.* 2005;58(6):715–25.
3. Templeton GF, Lewis BR, Snyder CA. Development of a measure for the organizational learning construct. *J Manag Inf Sys.* 2002;19(2):175–218.
4. Rashidi MM, Habibi M, Jafari Farsani J. The relationship between intellectual assets organizational learning capability at the institute for international energy studies. *Manag Hum Res Oil Ind.* 2010;11(4):59–76.
5. Hsu YH, Fang W. Intellectual capital and new product development performance: The mediating role of organizational learning capability. *Tech Foreca Soc Chan.* 2009;76(5):664–77.
6. Aghdasi M, Khakzar Bafruei A. Measuring level of organisational learning capabilities in hospitals. *I J Ind Eng Prod Manag.* 2009;19(4):71–8.
7. Bhatnagar J. Measuring organizational learning capability in Indian managers and establishing firm performance linkage: An empirical analysis. *Lear Org.* 2006;13(5):416–33.

8. Chiva R, Alegre J, Lapiedra R. Measuring organizational learning capability among the workforce. *Int J Manp*. 2007;28(3):224–42.
9. Easterby-Smith M, Crossan M, Nicolini D. Organizational learning: Debates past, present and future. *J Manag Stud*. 2000;37(6):783–96.
10. Alikhani M, Fazlollahtabar H. A mathematical model for optimizing organizational learning capability. *Adv Oper Res*. 2014;1:1–12.

Index

Annotation, 56, 58–59, 62
Articulation rules, 58, 60, 62
Asset, 19–20, 33, 46, 51, 65, 67, 94, 98, 122

Business, 3–5, 11, 19, 30, 33, 45, 47–49, 51–53, 55, 58, 63, 65–66, 84, 95–96, 99, 111, 117
Business environment, 19, 30, 84

Capability, 8, 19, 23–24, 33, 35, 37, 39, 40–43, 85, 93, 95, 111, 113–115, 117–122
Change, 13, 15–16, 19, 51, 66, 93, 95, 113–114, 116, 119
Clustering, 92, 111, 118, 121–122
Communication, 1–5, 19–22, 30, 42, 47, 62, 67, 92, 94, 97, 99, 100, 102, 104, 106, 109–110, 115–116, 119
Creativity, 92, 116
Culture, 1, 10, 67–68, 70, 72, 93–101, 103, 105, 107–111, 113, 115–117, 120
Customer, 3, 22, 83, 93, 111

Development, 2, 5, 7, 17, 20, 30, 35, 37, 65–66, 84, 87–88, 92, 95, 98–99, 114–116, 122
Dimension, 2, 4, 21, 33, 35, 38, 42, 56, 94–95, 97–100, 109–110, 113–121
Dynamic capabilities, 33, 35–37, 39, 41–42

Economic, 43, 65–69, 72, 79–80, 84, 115
Enterprise, 5, 22, 33, 42, 83, 85–92, 116
External knowledge, 8, 12, 19, 21, 29–31, 33, 35, 39

Feedback, 42, 56, 58, 62

Global learning, 1, 3, 5

Human resource, 3–4, 45, 47, 84, 86, 92, 95–96, 98
Hypothesis, 9, 12, 14, 21–22, 25, 40–41, 97, 102–108, 110

Information, 10, 12–13, 20–23, 29–31, 36–38, 46–47, 55–56, 58–59, 62–63, 66–70, 80, 86–89, 91–95, 97–100, 102, 113, 116, 119
Innovation, 1, 7–9, 11–17, 19–27, 29–31, 33, 35–37, 41–43, 80, 83, 85, 87, 89, 92–93, 95, 98, 100, 102, 104, 106–107, 109–111, 113, 119, 122
Innovation process, 7, 17, 19, 21, 29, 36, 122
Internal knowledge, 8, 23
Interoperability, 55, 58
Interval, 70–73, 76–78, 81

Knowledge acquisition, 8–9, 12–16, 33, 35, 37, 39–42, 113
Knowledge-based, 12, 33, 35, 37, 39, 53, 58, 60, 67, 83, 94
Knowledge continuity, 47–48, 53
Knowledge creation, 7–8, 12–17, 33, 42, 67, 93–94, 111
Knowledge development program, 87–88, 92
Knowledge intensity, 14–16
Knowledge management, 6–9, 11–12, 15, 17, 31, 51–52, 63, 67, 80, 83, 92–93, 96–100, 110–111, 117–118, 121
Knowledge process, 7–15, 17, 31
Knowledge retention, 45–47, 49–52, 93
Knowledge sharing, 5, 8–9, 12–17, 37, 55–58, 60–63, 65, 67–73, 75, 78–81, 83–90, 92–105, 107–114
Knowledge stock, 1
Knowledge transfer, 1–3, 5–6, 20, 45–47, 49, 66–68, 80, 84, 94, 111, 113–114, 116–117, 120

Learning capability, 113–115, 117–122
Long-term, 66, 70, 115
Loss, 45, 47–49, 51–53, 66, 70, 75–76, 78, 80

Macroeconomic, 66, 80
Manufacturing, 2–5, 20, 30, 37–39, 42, 101, 111
Mathematical model, 66
Measurement, 12, 22–25, 39, 85, 93, 111, 114, 118–122

Network, 11, 19, 21, 30–31, 33, 35–36, 38–39, 41–43, 51, 66, 80, 85, 87–92, 99

Organization, 7–8, 11, 13, 17, 19, 21, 23, 30–31, 33, 42–43, 45–53, 56, 62–63, 67–68, 80, 83–86, 92–111, 113–123
Organizational culture, 67, 93, 95, 96–97, 99, 101, 103, 105, 107, 109, 110–111, 117
Organizational learning, 63, 92, 113–117, 120–122

Performance appraisal, 3, 100, 102, 104–105, 109–110
Product innovation, 7, 37, 122

Risk, 45–48, 51, 65, 70, 72, 75, 78–80, 113–116, 119

Semantic web, 55–58, 60–63
Semi-structured knowledge, 56, 60, 62–63
Small and medium-sized enterprises (SMEs), 22
Statistical analysis, 23, 108
Strategic management, 20, 35, 42
Success factors, 47, 96–98, 111
Supply chain, 55–56, 60, 62–63

Tax payment, 65, 67–69, 71, 73, 75, 77, 79, 80–81
Team, 2, 39, 47, 68, 80, 92, 94, 95, 97–101, 104–105, 109–110, 113, 115–121
Technology, 21–22, 25, 27, 29–30, 43, 55–56, 67, 89, 92–93, 96, 110, 117

Uncertainty, 56, 65, 71–72, 114

Kolibri 360 GmbH
Ferdinand-Jühlke-Straße 7
99095 Erfurt, Deutschland
produktsicherheit@kolibri360.de